THE AIOS SERIES

DESIGN. BUILD. RUN. TRUST.

BUILDING YOUR FIRST AIOS

A BEGINNER'S GUIDE TO CREATING A PERSONAL AI OPERATING SYSTEM

FREDDIE POKRZYWA

BOOK ONE

3C PUBLICATIONS, LLC

Building Your First AIOS

A Beginner's Guide to Creating a Personal AI Operating System

This book was authored by Freddie Pokrzywa. Artificial intelligence was used as an editorial and writing assistant during the preparation of this manuscript. All technical content, architectural designs, implementation details, and opinions are the author's own and are based on the documented development of the AI Operating System described throughout this book.

First Edition

ISBN 979-8-9976875-0-2

3C PUBLICATIONS, LLC

Building the future, one system at a time.

DEDICATION

To my wife, Dorothy, my daughters, Cailey and Ashleigh, and of course, Bean.

This book is about teaching machines to observe, learn, and grow.

You taught me to do the same.

Dorothy, thank you for standing beside me through every late night, every ambitious idea, and every moment when the finish line felt impossibly far away. You've celebrated every success, carried me through every setback, and somehow always believed in me more than I believed in myself. Your love, patience, encouragement, and unwavering faith have been the foundation beneath every accomplishment in my life.

Cailey and Ashleigh, watching you grow has been the greatest privilege and joy of my life. I hope you always hold on to your curiosity, continue asking questions, and never lose the courage to build something that doesn't yet exist. The world will continue to change, technology will continue to evolve, but integrity, compassion, perseverance, and a willingness to keep learning will always matter more than any machine we create.

And Bean, thank you for being the quiet companion who is always reminding me to take a break, and proving that sometimes the best support doesn't need words.

No achievement, no career milestone, and no book will ever be greater than being your husband, your dad, and yes, Bean's favorite human.

This book, and everything that follows, is dedicated to you.

With all my love,

Freddie

HOW THIS BOOK WAS WRITTEN

This book was written by me, with the assistance of artificial intelligence.

Over the years throughout my IT career, I've documented my work through hundreds of Markdown files, architecture documents, design notes, implementation journals, diagrams, and lessons learned while building systems. Capturing not only **what** I built, but **why** I built it that way has become a habit and an invaluable reference throughout my career.

When I began building my personal AI Operating System, **Cora**, I approached it the same way. Every architectural decision, experiment, success, setback, and lesson learned was documented. Those engineering journals became the foundation for this book.

Rather than starting with a blank page, I used AI as a writing partner. It helped me organize ideas, improve clarity, refine explanations, identify inconsistencies, and transform years of engineering notes into something I hope is easier to read and learn from. It did **not** replace the engineering, the architecture, or the experiences described here.

Every concept, design decision, code example, success, mistake, and lesson in this book comes from building the system myself. When AI suggested improvements, I evaluated them the same way I would feedback from another engineer: if they improved the book while remaining technically accurate, I used them. If they didn't, I rewrote or discarded them.

I believe this is how AI should be used: **not as a replacement for expertise, but as a tool that amplifies it.**

My goal is simple: to share what I've learned honestly, transparently, and in a way that helps you build something of your own.

> **One final confession:** AI helped organize my thoughts into something humans can read. Otherwise, you'd probably be reading three architecture diagrams, a ServiceNow tangent, and wondering how we got there. The engineering is mine. The clearer writing is thanks to AI. **Any bugs in the code? Those are almost certainly my contribution.**

— **Freddie Pokrzywa**

ACKNOWLEDGEMENTS

My sincere thanks to the first readers of this book, whose invaluable feedback helped refine and improve the material:

- Bryan Crowder
- Gillian Adams
- Nachiket Suryawanshi
- Phanom Chanthavong
- Peter Wilkinson
- Duc Haba

CONTENTS

What's in this book

Preface

> *Building Your First AIOS: A Beginner's Guide to Creating a Personal AI Operating System*

A small scene

Imagine with you open a page. A soft orb breathes in the center of a dark, star-flecked field. You say, out loud, "Pip, show me my inbox and switch to bubble view." The orb brightens. A panel slides open on the left; the cards on screen dissolve into floating bubbles; one of them, the inbox, drifts to the front. You never touched the mouse.

Then you type: "Actually, always start me in bubble view." Pip replies, "Got it, I'll remember that," and it does. Tomorrow, and every day after, it opens the way you asked.

Nothing here is magic, and nothing here is large. The whole system is a single HTML file, a few hundred lines of Python, and a database with a handful of tables. You are going to build all of it, and by the end you will understand every line.

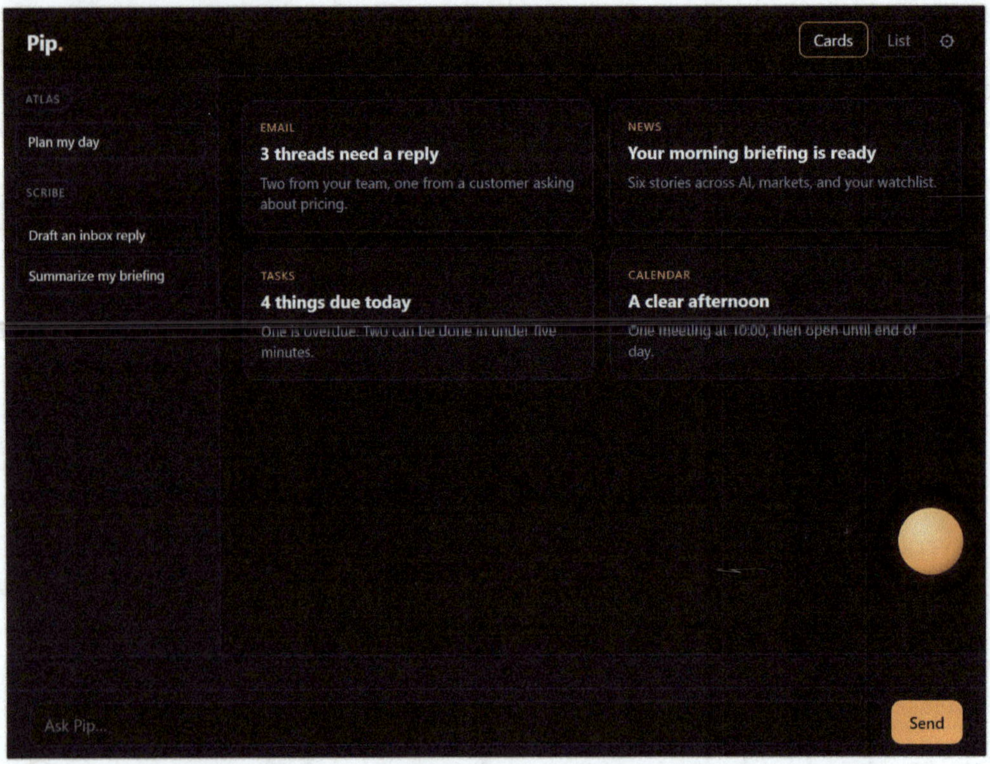

FIGURE P.1 *The finished Pip, in its default cards view. It looks like an app; the difference is that the AI can see and drive every part of it. Each chapter adds one more piece of this screen.*

WHY

While designing and building "Cora" my local AI Operating System to help with my personal and work life, I kept detailed notes as to what worked and what didnt, and the whys. I stored each of these sessions in roughly 154 markdown files. The plan was to build a final architecture document using AI to document the journey and areas to look out for. When talking about Cora to friends, family, and coworkers a fair bit of them wanted to know the "how to's" for each piece. So I decided to share this information it might help to put all the data I have colleted and create something that folks could follow and build a base version for themselves. The folks were a mixed variation of very technical to some what technical but dabbled with coding.

The WHY NOW

I have been in IT for the past 30+ years and the last 14 in the ServiceNow ecosystem. People often ask why I have not written about IT in the past, and what inspired me to write about artificial intelligence. The truth is, it wasn't the technology that inspired me. It was the future. A future where the tools we create help people spend more time doing what matters most with the people they love. So, every page was written with the hope that it might help someone do just that.

The one idea

Most people, when they imagine "adding AI" to an application, picture a chat box in the corner. Text goes in, text comes out, and the AI is a clever stranger who can talk about your app but can't touch it.

This book is built on the opposite idea:

> **An AIOS is not a chatbot with a UI bolted on. It is a UI that an AI can perceive and drive.**

Read that twice, because everything else follows from it. In an ordinary app, the interface is for the human and the AI is a bystander. In an AI Operating System, the interface is a **shared workspace**: the human and the AI both look at the same screen, and both can change it. The AI is no longer a stranger describing your app from the outside; it is a resident who can see the room and move the furniture.

To make that real, you need to build exactly three things. They are the spine of this entire book:

1. **Perception:** the AI can *see* the current state of the screen.
2. **Control:** the AI can *act* on the screen, through a safe and limited set of moves.
3. **Self-awareness:** the AI *knows what it is*, including its own preferences and its own architecture.

Get those three loops working and you have an operating system, not an app. Everything else (voice, background agents, live updates, a breathing orb) is decoration on that foundation. It is beautiful, worthwhile decoration, but it is still decoration.

Who this book is for

You'll feel at home here if you can write a little HTML, a little JavaScript, and follow a Python file without panicking. You do **not** need to have built a backend before, used a database before, or touched an AI API before. Where those things come up, we build them from zero and explain why.

You will get the most out of this book if you are the kind of person who is not satisfied by a library that does the magic for you. There are frameworks that will hand you an "AI agent" in five lines. This is not that. This is the version where you build the machine yourself, out of small parts you can name, so that when it misbehaves, and it will, you know exactly where to look.

If you have never programmed at all, this book will move too fast. Come back after a first course in web development. If you are a senior engineer, some early chapters will feel gentle; skim them, but don't skip the **gotcha** sections, because that's where the hard-won specifics live, and they're the same for everyone.

What you'll need

- A computer with **Python 3.11 or newer** and a terminal you're willing to type into.
- A text editor. Anything. VS Code if you have no opinion.
- **A brain for Pip.** You have two choices, and Pip is built to accept either:
    - An **Anthropic API key** (cloud, higher quality, costs a little per message), or
    - Ollama installed locally (free, private, runs on your own machine).

  Chapter 2 sets this up. If you're not sure, start with Ollama; it costs nothing and never leaves your laptop.

That's the entire shopping list. No accounts to create beyond the one for your chosen brain, no cloud to deploy to, no build tools to fight.

How the book is built

We grow **one application, slowly.** Chapter 3 gives Pip a face. Chapter 4 gives it a memory. Chapter 5 gives it a brain. Chapter 6 gives it eyes, Chapter 7 gives it hands, Chapter 9 gives it a sense of self. You are never starting over; each chapter opens the same project folder and adds to it. By the last page, the toy from Chapter 3 has become the small scene that opened this preface.

Every build chapter moves through the same four beats, and once you feel the rhythm you'll never be lost:

> **Show the seam → build the dumb version → add the AI → show the gotcha.**

First we name the *seam*, the boundary we're about to build across, so you understand the point before the syntax. Then we build the **dumb version**: the simplest thing that works, with no intelligence in it at all, because if you can't drive the interface by hand, the AI won't be able to either. Then we **add the AI**, plugging the model into the seam we just built. And finally the **gotcha**: the specific, unglamorous thing that will break and cost you an afternoon if nobody warns you. Those gotcha sections are the reason this book exists. Anyone can show you the happy path.

The north star

Throughout the book you'll see sidebars marked **North star · Cora**. Cora is a full, production-scale AIOS, the same shape as Pip but industrial: a real Postgres database, seven specialized agents, a live-updating pipeline, voice, a WebGL background. Pip is the version you can hold in your head; Cora is the version that shows you where the road leads.

You never need the north-star sidebars to build Pip. They're there for the moment you finish a chapter, look at your 40 lines of working code, and wonder: *"okay, but how would this survive a thousand users and a dozen features?"* The answer is in the sidebar. Take it or leave it; Pip runs fine either way.

A promise, and a warning

The promise: if you type along, chapter by chapter, you will end up with a running AI Operating System that you built and understand, and, more valuable than the code, a clear mental model of *how these systems actually work* underneath the hype. You will never again look at an "AI that can use your app" and wonder what's inside. You'll know. You'll have built one.

The warning: this is real software, and real software has warts. Models drop your instructions silently. Regexes fail on a missing newline. Hot-reload hangs when background tasks pile up. We don't hide any of this. It goes in the gotcha sections, because pretending the happy path is the whole story is how tutorials set you up to fail. Expect friction. The friction is where the learning is.

How to begin

Turn to **Chapter 1: The Mental Model.** It's short and has no code; it just makes sure the three loops are firmly in your head before your hands start moving. Then Chapter 2 sets up your workshop, and from Chapter 3 on, you build.

One app. Three loops. Fourteen chapters.

Let's give it eyes.

ORIENTATION

How to Read This Book

Read it **in order.** Each chapter adds working code to the same Pip project; skipping ahead leaves you with a codebase that won't run. Every build chapter follows the same four beats, so you always know where you are:

> **Show the seam → build the dumb version → add the AI → show the gotcha.**

- **The seam** is the boundary you're about to build across, named before any code so you understand *why* before *how*.
- **The dumb version** is the simplest thing that works, with no intelligence in it yet.
- **Add the AI** is where the model plugs into the seam you just built.
- **The gotcha** is the thing that will bite you, the pages tutorials usually hide.

Two recurring devices:

- **North star** (Cora) sidebars show how a production AIOS handles the same problem at scale. You never *need* them to build Pip; they're the "where this goes next" view.
- **Exercises** at the end of each build chapter extend Pip in a small way. Do them; they're where the understanding sets.

Code you should type into a file is shown in labeled blocks. Commands you run in a terminal are shown like this:

```bash
python -m uvicorn server:app --reload --port 8000
```

CHAPTER ONE

The Mental Model

> *No code in this chapter. We're building the picture in your head that every later chapter fills in. Read it once now; you'll come back to it.*

There's a temptation, when learning something hands-on, to skip straight to the keyboard. Resist it for twenty minutes. The single most common reason people build a confused, fragile AIOS is not a bug. It's that they never got clear on what the thing *is*. Once the mental model is solid, the code is almost boring. Let's make it solid.

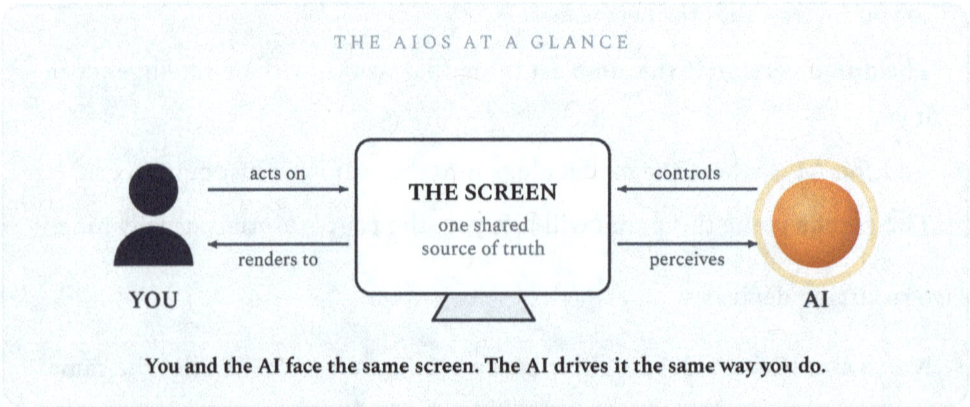

FIGURE 1.1 *Level 0 · the master architecture, no backchannel and no hidden state.*

Three things people call "AI in an app"

The phrase "AI-powered app" covers three genuinely different architectures. They are not points on a spectrum of *quality*. They are different *shapes*. Confusing them is the root of most bad designs.

The chatbot

Text goes in, text comes out.

You type a question, the model answers. It might be a brilliant answer. But the model has no hands and no eyes: it cannot see anything on your screen, and it cannot change anything on your screen. It is a very smart pen-pal who happens to live in a box on your page. When you close the box, nothing about your app is different.

Almost every "add AI to your product" tutorial builds this. It's useful, and it's the easy 80%. It is also *not an operating system*, and it never becomes one by getting a bigger model.

The agent

Text goes in, **actions** come out.

This is a real step up. An agent doesn't just talk; it can *do* things. You give it tools (functions it's allowed to call: search the web, query a database, send an email), and it decides which to call and with what arguments. Modern models are good at this. An agent can book a flight or file a ticket.

But notice what's still missing: an agent is usually **blind to your interface.** It knows about its tools, not about your screen. It can call `send_email(...)`, but it has no idea that the user is currently staring at a half-composed draft in the corner of the page. The agent acts *through* an API, not *on* a shared surface that the human is also looking at. It has hands, but it's working in the dark.

The AIOS

Text goes in, **and the AI can see the screen and drive it**, the same screen the human sees and drives.

This is the shape this book builds. The interface is not *for the human, described to the AI*. The interface is a **shared workspace** that both of them read from and write to. The AI can see that a panel is open, that you're in bubble view, that the inbox card is at the front. And it can change any of those, not by some secret backchannel, but by doing the same things you could do by clicking.

That's the whole difference, and it's worth saying plainly:

	SEES YOUR SCREEN?	CHANGES YOUR SCREEN?
Chatbot	No	No
Agent	No	Through tools, in the dark
AIOS	Yes	**Yes, the same controls you have**

An AIOS is an agent that has been given *eyes on the actual interface* and *hands on the actual controls.* That is the entire idea. The rest of this book is just: how do you build the eyes, and how do you build the hands, safely.

The loop

Here's the picture to keep in your head. There are three actors (you, the screen, and the AI) and they pass control around in a cycle:

```
┌───────────────────────────────────────────────────────┐
│                                                       │
▼                                                       │
YOU ──acts──▶ SCREEN STATE ──perceives──▶ AI ──controls──┐
▲             (one source                                │
│              of truth)  ◀──────────────────────────────┘
└──renders───────┘
```

Trace it slowly:

1. **You act.** You click, type, or speak. This changes the **screen state.**
2. **The screen renders** back to you, so you see the result. (Ordinary app, so far.)
3. **The AI perceives** the screen state. It reads a description of what's currently on screen.
4. **The AI controls** the screen. It issues an action, which changes the screen state.
5. ...which renders back to you, and the cycle continues.

The important thing about this diagram is that **it has no fixed starting point.** You and the AI are both writing to the same state, and both reading from it. Sometimes you move first and the AI reacts; sometimes the AI moves and you react. Neither is "the driver" and the other "the passenger." They share the wheel.

And notice: the AI never reaches *around* the interface to some hidden mechanism. Every arrow the AI draws goes *through* the screen state, the exact same state you manipulate. That constraint is not a limitation to work around. It's the safety property that makes the whole thing sane, and we'll lean on it hard in Chapter 7.

The three loops, located on the diagram

The three things you're building map directly onto arrows in that picture:

- **Perception** is the `SCREEN STATE —perceives—▶ AI` arrow. Building it (Chapter 6) means assembling an honest, current description of the screen and handing it to the model on every turn. A model with a broken perception arrow *guesses* about your app; a model with a good one *knows.*

- **Control** is the `AI —controls—▶ SCREEN STATE` arrow. Building it (Chapter 7) means defining a small, fixed vocabulary of moves the AI is allowed to make, and wiring each one to something that actually changes the screen. This is the highest-leverage and highest-risk arrow, which is why it gets the most careful chapter.

- **Self-awareness** is a smaller loop that lives *inside* the AI node: the model knowing its own remembered preferences and its own architecture. Building it (Chapter 9) means feeding the AI, every turn, a little description of *itself*: what it's been asked to remember, and how it's built. It's the arrow that turns a tool into something that feels like a system.

Everything else in the book (the data layer, the brain, skills, agents, live updates, voice) exists to make those three arrows real, fast, and pleasant.

The one rule that makes it all work

If you take a single sentence from this chapter into the rest of the book, take this one:

> **State lives in one place, readable by both the UI and the AI.**

Look back at the loop. It only works if "screen state" is a single, real, shared thing. If the truth about "what view am I in" lives in three different places, a JavaScript variable here, a CSS class there, a database column somewhere else, then the AI's perception can't be trusted (it read one copy while another changed) and the AI's control can't be trusted (it changed one copy while the screen reads another). The loop falls apart into a mess of race conditions and lies.

So from Chapter 3 onward, we are disciplined about this. Pip will keep its state in exactly two well-defined places:

- **On the page itself**, as `data-*` attributes on the root `<html>` element, readable by CSS *and* JavaScript *and* describable to the AI. ("What view are we in?" is `<html data-view="bubbles">`. One place. No arguments.)
- **In the database**, for anything that must persist or that the AI needs to change durably: the cards, the tasks, the remembered preferences.

When there is one source of truth, the AI can change what you see simply by changing that truth. That is the deep reason an AIOS is *possible*, and the deep reason a sloppy one is *impossible*. Guard this rule and an AIOS becomes a pleasant series of small additions. Break it and every feature you add will fight the last one.

NORTH STAR · CORA

Cora holds exactly this line, at scale. Its screen state lives on `<html>` as data attributes (`data-view`, `data-theme`, `data-focus`, and more), and its durable content lives in Postgres. When Cora's AI decides to switch you to bubble view, it doesn't run some special animation routine. It sets `data-view="bubbles"`, the *same attribute* the view-selector buttons set when you click them. The AI has no power the user doesn't; it just has a description of the room and a hand on the same switches. Every one of Cora's thirteen AI actions bottoms out in "do a thing the user could already do." That isn't an accident of implementation. It's the mental model from this chapter, enforced in code.

What you now know

- The three shapes (chatbot, agent, AIOS) and that the difference is *eyes and hands on the shared interface*, not model size.
- The loop: you and the AI both read and write one shared screen state, passing control around a cycle with no fixed start.
- The three loops you'll build (perception, control, self-awareness) and where each lives on that diagram.
- The one rule: **one source of truth, readable by both sides.**

Before you turn the page

You don't write any code for these; they're just to check the model took hold. Answer them in your head (or a notebook):

1. A weather app has a chat box where you can ask "will it rain tomorrow?" Is it a chatbot, an agent, or an AIOS? What single capability would you have to add to move it up one level?
2. In the loop diagram, why is it a *problem* if the AI could change the screen through some path that *doesn't* go through the shared screen state? Name one concrete thing that could go wrong.
3. Pick any app you use daily. Where does its "what am I currently looking at" state live? Is it in one place, or scattered? Could an AI honestly perceive it?

Hold those answers loosely. Chapter 2 sets up your workshop (Python, the project folder, and a server that says hello) and Chapter 3 starts turning this picture into a running thing.

CHAPTER TWO

Set Up Your Workshop

> *This is the first chapter with code. By the end you'll have a folder called* `pip/`*, a Python server that answers a request, a page it serves, and a brain waiting to be wired in. Every later chapter opens this same folder and adds to it.*

Building software is mostly building the *place* you build software. A carpenter spends the first hour of a new job setting up sawhorses. We're doing the same: this chapter has no AI in it at all, because the AI has nothing to plug into until the workshop exists. Resist the urge to rush. A clean setup now saves you a dozen confusing errors later.

Here's the whole plan for this chapter:

1. Confirm Python is installed.
2. Make the `pip/` project folder and an isolated environment inside it.
3. Install two libraries.
4. Write a tiny server and a page for it to serve.
5. Run it and see it in a browser.
6. Set up a **brain**, either local (Ollama, free) or cloud (Anthropic), ready for Chapter 5.
7. Put it all under version control so nothing is ever lost.

WHERE CHAPTER 2 ENDS

pip/

.venv/	isolated Python sandbox · git-ignored
.env	config + secrets · git-ignored
.gitignore	tells git what to skip
requirements.txt	the two libraries you installed
server.py	FastAPI: serves the page + the API
index.html	the page (a placeholder, for now)

FIGURE 2.1 *The finished workshop. Seven items, and everything later hangs off them.*

1. Confirm Python

Pip's backend is Python. You need **version 3.11 or newer.** Open a terminal and check:

```bash
python --version
```

If that prints `Python 3.11.x` or higher, you're set. If it prints a 3.10 or lower, or says the command isn't found, try `python3 --version` instead. On macOS and Linux the command is often `python3`.

> **Gotcha: `python` vs `python3`**. On Windows, `python` usually works. On macOS/Linux, `python` may point at an ancient Python 2 or nothing at all, and `python3` is the real one. Whichever name prints 3.11+ is the one you'll use for the rest of this book. If you have to type `python3`, mentally substitute it everywhere I write `python`.

If Python is missing or too old, install it from python.org/downloads (on Windows, tick "**Add Python to PATH**" in the installer; it saves a headache) or via your package manager (`brew install python` on macOS, `sudo apt install python3 python3-venv` on Debian/Ubuntu). Then re-check the version.

2. Make the project folder and a virtual environment

Pick a place for your projects and create the folder:

```bash
mkdir pip
cd pip
```

Everything from here on happens *inside* `pip/` .

Now create a **virtual environment**, an isolated Python sandbox that lives in this folder, so the libraries Pip needs don't collide with anything else on your machine:

```bash
python -m venv .venv
```

That makes a hidden `.venv/` folder. Now **activate** it:

- **macOS / Linux:**

  ```bash
  source .venv/bin/activate
  ```

- **Windows (PowerShell):**

  ```bash
  .venv\Scripts\Activate.ps1
  ```

- **Windows (classic `cmd`):**

  ```bash
  .venv\Scripts\activate.bat
  ```

When it's active, your prompt gets a `(.venv)` prefix. That prefix is how you know you're working inside Pip's sandbox.

> **Gotcha: Windows won't run the activate script.** PowerShell may refuse with *"running scripts is disabled on this system."* That's a one-time security setting. Fix it by running this once, then activating again:
>
> ```bash
> Set-ExecutionPolicy -Scope CurrentUser RemoteSigned
> ```

> **Gotcha: you'll need to re-activate every session.** The virtual environment isn't "on" forever. Every new terminal you open to work on Pip, `cd` into `pip/` and run the activate command again. If `pip install` ever seems to install into the wrong place, or an import fails that you *know* you installed, check for the `(.venv)` prefix first. It's almost always a forgotten activation.

3. Install the libraries

With the environment active, install the two things Pip needs to serve a page and an API:

```bash
pip install "fastapi[standard]" uvicorn
```

- **FastAPI** is the web framework. It turns Python functions into web endpoints.
- **Uvicorn** is the server that runs FastAPI and listens for browser requests.

Record what you installed in a `requirements.txt` file, so anyone (including future you, on a new machine) can recreate the environment exactly:

`pip/requirements.txt`

```
fastapi[standard]
uvicorn
```

We'll add a line or two to this file as later chapters pull in more libraries. For now, two.

4. Write the server and a page

Two files. First, the server:

`pip/server.py`

```python
from pathlib import Path

from fastapi import FastAPI
from fastapi.responses import FileResponse

# The folder this file lives in, so we can find index.html no matter
# what directory the server was started from.
BASE = Path(__file__).resolve().parent

app = FastAPI(title="Pip")


@app.get("/api/health")
def health():
    """A simple endpoint to prove the API is alive."""
    return {"status": "ok", "app": "pip"}


@app.get("/")
def index():
    """Serve the single-page frontend."""
    return FileResponse(BASE / "index.html")
```

Read what's there, because this shape recurs for the rest of the book:

- `app = FastAPI(...)` creates the application.
- `@app.get("/api/health")` says: *when a browser asks for* `/api/health`*, run this function and send back whatever it returns.* Returning a Python dict makes FastAPI send JSON. This is the seed of every `/api/...` endpoint we'll build.
- `@app.get("/")` serves the page itself, as a file. **One process serves both the page and the API.** That's the "single-process" promise from the preface, and it's why an AIOS this small has no separate frontend server to manage.

Now the page. It's a placeholder, and Chapter 3 replaces it with Pip's real face, but it proves the whole pipe works:

`pip/index.html`

```html
<!doctype html>
<html lang="en">
<head>
  <meta charset="utf-8">
  <meta name="viewport" content="width=device-width, initial-scale=1">
  <title>Pip</title>
  <style>
    body { font-family: system-ui, sans-serif; background: #0e1117;
color: #e6e9ef;
           display: grid; place-items: center; height: 100vh; margin: 0;
}
    .card { text-align: center; }
    h1 { font-size: 3rem; margin: 0; }
    p  { color: #8b93a7; }
    code { color: #fb923c; }
  </style>
</head>
<body>
  <div class="card">
    <h1>Pip</h1>
    <p>Your AI operating system is awake.</p>
    <p>Health check: <code id="health">checking…</code></p>
  </div>
  <script>
```

```
    // Ask our own API if it's alive, and show the answer.
    fetch('/api/health')
      .then(r => r.json())
      .then(data => { document.getElementById('health').textContent =
JSON.stringify(data); })
      .catch(() => { document.getElementById('health').textContent =
'unreachable'; });
  </script>
</body>
</html>
```

That page does one clever thing: on load, its script calls `/api/health`, the endpoint you just wrote, and prints the answer. If it works, you'll *see the frontend and the backend talking to each other.* That handshake is the entire architecture in miniature.

5. Run it

From inside `pip/`, with the environment active:

```bash
python -m uvicorn server:app --reload --port 8000
```

Reading the command: run uvicorn, load the `app` object from `server.py` (`server:app`), watch the files and restart on any change (`--reload`), and listen on port 8000. You should see a line like `Uvicorn running on http://127.0.0.1:8000` .

Open **http://127.0.0.1:8000** in your browser. You should see "Pip: Your AI operating system is awake," and the health line should read `{"status":"ok","app":"pip"}` . If it does: the server is serving the page, the page is calling the API, and the API is answering. The pipe is complete.

Leave this running in its own terminal while you work. Because of `--reload` it restarts automatically whenever you save a file, so you rarely need to stop it.

> **Gotcha: `uvicorn: command not found`.** Always start it with `python -m uvicorn ...`, not a bare `uvicorn ...`. The `python -m` form guarantees you're using the uvicorn installed in *this* environment, not some other one (or none). Bare `uvicorn` works only when the environment's scripts are on your PATH, which isn't guaranteed.

> **Gotcha: "address already in use."** Port 8000 is taken, usually by an earlier copy of this same server you forgot to stop. Either stop the old one (find its terminal and press `Ctrl+C`) or run this one on a different port with `--port 8001` and visit that instead.

6. Give Pip a brain

Pip can serve a page, but it can't *think* yet; that's Chapter 5. What we do now is install the brain and get it ready, so Chapter 5 is pure wiring. You have two choices. **You only need one.**

Option A: Ollama (local, free, private), recommended to start

Ollama runs a language model entirely on your own machine. It costs nothing, needs no account, and never sends your data anywhere.

1. Download and install it from ollama.com/download.
2. Pull a small, capable model:

```bash
ollama pull llama3.2
```

   (`llama3.2` is a 3-billion-parameter model, small enough to run on a laptop and smart enough to drive Pip. If your machine is tight on memory, `ollama pull llama3.2:1b` is even smaller.)

3. Confirm it works:

```bash
ollama run llama3.2 "Say hello in five words."
```

   If it answers, your brain is installed. Ollama runs a little server in the background at `http://localhost:11434`, and Chapter 5 will talk to it there.

Option B: Anthropic (cloud, paid, higher quality)

If you'd rather use a top-tier cloud model, sign up at console.anthropic.com, create an **API key**, and keep it somewhere safe for the next step. Cloud models cost a small amount per message and send your prompts to Anthropic's servers. We'll pick the exact model in Chapter 5; the setup here is just having a key.

> **Prohibited-action note:** this book never asks you to paste an API key into a web page or a form. A key is a credential. It goes in one place only, the local `.env` file below, which stays on your machine and is never committed to version control. If any site or tool ever asks you to type your key into it, stop and be suspicious.

Record your choice in `.env`

Create a file named `.env` (just the extension, no name) in `pip/`. This holds Pip's configuration and secrets. It stays local, and we'll make sure it's never committed.

pip/.env

```
# Which brain Pip uses: "ollama" (local, free) or "anthropic" (cloud).
LLM_PROVIDER=ollama

# --- Ollama settings (used when LLM_PROVIDER=ollama) ---
OLLAMA_BASE_URL=http://localhost:11434
OLLAMA_MODEL=llama3.2

# --- Anthropic settings (used when LLM_PROVIDER=anthropic) ---
# Paste your key here if you chose Option B. Leave blank otherwise.
ANTHROPIC_API_KEY=
# The model is chosen in Chapter 5.
```

If you chose Ollama, you're done, because the defaults are correct. If you chose Anthropic, set `LLM_PROVIDER=anthropic` and paste your key after `ANTHROPIC_API_KEY=`.

7. Put it under version control

An AIOS changes fast. Version control is your undo button for the whole project. Initialize a repository and tell it what to ignore:

pip/.gitignore

```
.venv/
__pycache__/
*.pyc
.env
pip.db
```

That `.env` line is the important one: **it keeps your secrets and local config out of the repository.** The `pip.db` line ignores the database file we'll create in Chapter 4. Now initialize:

```bash
git init
git add .
git commit -m "Chapter 2: Pip serves a page and answers a health check"
```

> **Gotcha: check `.env` really is ignored.** Run `git status`. You should *not* see `.env` in the list of files to be committed. If you do, your `.gitignore` isn't being picked up (wrong folder, or a typo). Fix it before you ever commit, because a key committed to git is a key you have to treat as leaked, even after you delete it, since it lives in the history.

We'll cover the full branch-and-changelog workflow in Chapter 12. For now, one commit per chapter is plenty.

Your workshop

Here's what `pip/` looks like now:

```
pip/
├── .venv/            ← your isolated Python sandbox (git-ignored)
├── .env              ← config + secrets (git-ignored)
├── .gitignore
├── requirements.txt
├── server.py         ← FastAPI: serves the page + the API
└── index.html        ← the page (a placeholder, for now)
```

Small. That's the point. Everything the rest of the book adds hangs off these files.

What you built

- A Python virtual environment and a two-line dependency list.
- A FastAPI server that serves a page **and** an API endpoint from one process.
- A page that proves the frontend and backend can talk.
- A brain, local or cloud, installed and configured, waiting for Chapter 5.
- A git repository with your secrets safely ignored.

Exercises

1. **Add a second endpoint.** Give `server.py` a `GET /api/hello` that returns `{"message": "Hi from Pip"}`. Save, and thanks to `--reload` just refresh, then visit `http://127.0.0.1:8000/api/hello` directly. You've now added an API route by yourself.
2. **Break it on purpose.** Stop the server and start it again with `--port 8000` twice, in two terminals. Read the error from the second one. Recognizing "address already in use" now means you won't lose five minutes to it later.
3. **Prove the ignore works.** Run `git status` and confirm `.env` is *not* listed. Then temporarily rename `.gitignore` to `.gitignore.off`, run `git status` again, and watch `.env` appear. Rename it back. Now you've seen exactly what `.gitignore` is doing for you.

Checklist before Chapter 3

- ☐ `python --version` shows 3.11 or newer.
- ☐ The `(.venv)` prefix appears in your prompt when you're working.
- ☐ `http://127.0.0.1:8000` shows the Pip page with a green-looking health line.
- ☐ You have either Ollama installed with a model pulled, **or** an Anthropic key in `.env`.
- ☐ `git status` does **not** list `.env`.

With the workshop built and a brain on standby, Chapter 3 begins the real work: replacing that placeholder page with Pip's actual interface, rendered from data so the AI can drive it later.

CHAPTER THREE

Build the Shell

> *Now Pip gets a face. By the end of this chapter it has a real interface, with a header, a set of cards, a view toggle, and a detail modal, and, crucially, all of it is rendered from data rather than written into the HTML by hand. There's still no AI. That's deliberate: we build the thing the AI will drive before we let it drive.*

Think back to the loop from Chapter 1. The AI perceives the screen and acts on it, but "the screen" has to *exist*, and it has to be the kind of screen an AI can address. This chapter builds exactly that. Everything we do here answers one question in advance: *when the AI wants to change what the user sees, what will it actually reach for?* The answer, every time, will be "a piece of data or a `data-*` attribute," never "a chunk of hand-written HTML."

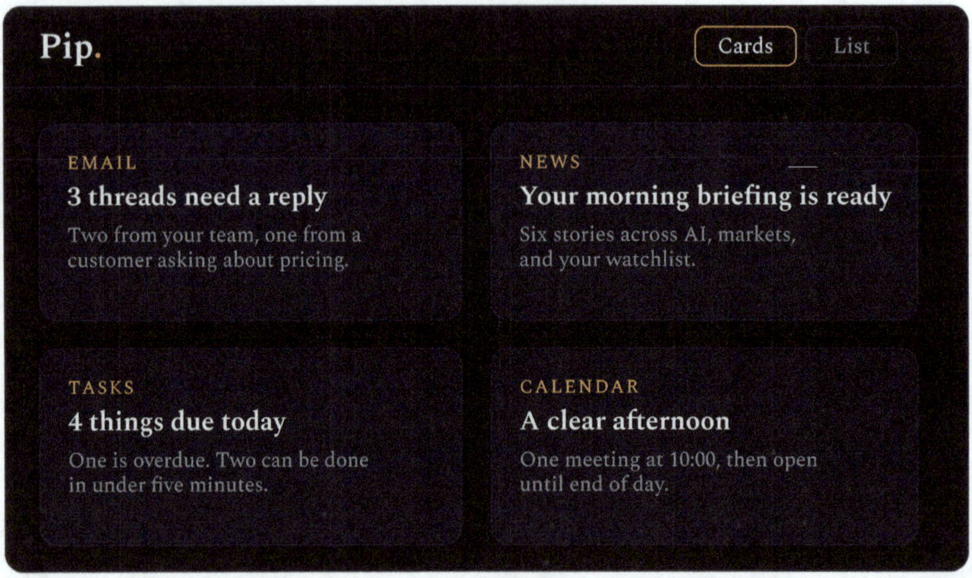

FIGURE 3.1 *Pip's shell. Four cards, a view toggle, a detail modal, all rendered from data, no AI yet.*

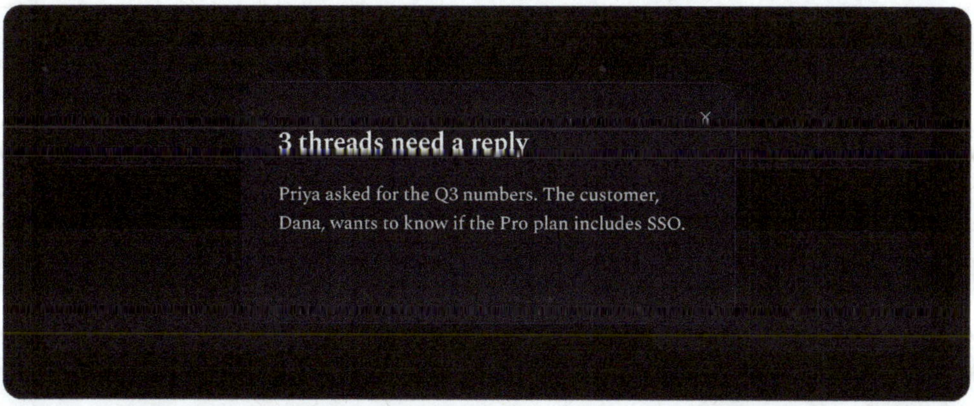

FIGURE 3.2 *The detail modal, drawn from the card's own data, not hand-written into the page.*

The seam

> **The interface exists before any intelligence does. If you can't drive it by hand, the AI can't either.**

So we build a fully working interface with zero AI in it. A human can click the view toggle, open a card, close it. Later, in Chapter 7, the AI will click those same things, through the same mechanisms. If we build the interface so that *only a human clicking a mouse* can operate it, we'll have to tear it apart later to let the AI in. Build it right the first time and the AI slots in with almost no changes.

The one rule from Chapter 1 governs everything here:

> **State lives in one place, readable by both the UI and the AI.**

For Pip, "what am I looking at" state will live on the root `<html>` element as `data-*` attributes, and the *content* will come from the server. Two sources of truth, each clear about what it owns.

Build the dumb version

Step 1: serve the card data

Right now `server.py` serves a page and a health check. Give it one more endpoint: a list of **cards**. A card is a unit of information Pip shows the user, such as an inbox summary, a news brief, or a task list. For now the data is a hand-written Python list; Chapter 4 moves it into a real database. The frontend won't know the difference, and that's the point.

Replace `server.py` with this:

pip/server.py

```python
from pathlib import Path

from fastapi import FastAPI
from fastapi.staticfiles import StaticFiles

BASE = Path(__file__).resolve().parent

app = FastAPI(title="Pip")

# The cards Pip shows. Each has a `slug`: a short, unique, URL-safe id.
# The slug is the name the whole system uses to refer to this card:
# the DOM, the database (next chapter), and the AI (Chapter 7) all use
it.
CARDS = [
    {
        "slug": "inbox",
        "eyebrow": "Email",
        "title": "3 threads need a reply",
        "body": "Two from your team, one from a customer asking about
pricing.",
        "detail": "<p>Priya asked for the Q3 numbers. The customer, "
                  "Dana, wants to know if the Pro plan includes SSO.
</p>",
    },
    {
        "slug": "briefing",
```

```
        "eyebrow": "News",
        "title": "Your morning briefing is ready",
        "body": "Six stories across AI, markets, and your watchlist.",
        "detail": "<p>Top item: a new open-weights model just shipped. "
                  "Two of your watched stocks moved more than 3%.</p>",
    },
    {
        "slug": "tasks",
        "eyebrow": "Tasks",
        "title": "4 things due today",
        "body": "One is overdue. Two can be done in under five
minutes.",
        "detail": "<p>Overdue: send the invoice to Dana. "
                  "Quick wins: reply to Priya, book the meeting room.
</p>",
    },
    {
        "slug": "calendar",
        "eyebrow": "Calendar",
        "title": "A clear afternoon",
        "body": "One meeting at 10:00, then open until end of day.",
        "detail": "<p>10:00 standup. After that, nothing scheduled. "
                  "A good window for focused work.</p>",
    },
]


@app.get("/api/health")
def health():
    return {"status": "ok", "app": "pip"}


@app.get("/api/cards")
def list_cards():
    """Return every card as JSON."""
    return CARDS


# Serve the frontend files (index.html, bootstrap.js, shell.js) from
this folder.
# IMPORTANT: mount this AFTER the /api routes above, so /api/* is
matched first.
# `html=True` makes "/" serve index.html automatically.
```

```
app.mount("/", StaticFiles(directory=BASE, html=True), name="static")
```

Two things changed. We added `/api/cards`, and we swapped the hand-written `/` route for a **static mount**, one line that serves *every* file in the folder (`index.html` today, `bootstrap.js` and `shell.js` in a moment). The mount goes last, so the specific `/api/...` routes win before the catch-all takes over.

With the server running, visit **http://127.0.0.1:8000/api/cards** directly. You'll see the raw JSON array. That's the data the frontend is about to render.

Step 2: give the page a shape

Replace `index.html`. This version defines the *structure*, a header, an empty card layer, an empty modal, and the styling, but writes **no cards**. The cards will be filled in at runtime. Notice the empty `<main id="cardLayer">`: that emptiness is intentional and important.

`pip/index.html`

```html
<!doctype html>
<html lang="en" data-view="cards">
<head>
  <meta charset="utf-8">
  <meta name="viewport" content="width=device-width, initial-scale=1">
  <title>Pip</title>
  <style>
    :root { --bg:#0e1117; --panel:#161b22; --line:#262d3a;
            --text:#e6e9ef; --muted:#8b93a7; --accent:#fb923c; }
    * { box-sizing: border-box; }
    body { margin:0; font-family: system-ui, sans-serif;
           background: var(--bg); color: var(--text); }

    .topbar { display:flex; align-items:center; justify-content:space-
between;
              padding:16px 24px; border-bottom:1px solid var(--line); }
    .brand { font-weight:700; font-size:1.3rem; }
    .brand span { color: var(--accent); }
```

```
    .view-toggle button { background:transparent; border:1px solid var(-
-line);
        color:var(--muted); padding:6px 14px; border-radius:8px;
cursor:pointer;
        font: inherit; }
    .view-toggle button.active { color:var(--text); border-color:var(--
accent); }

    /* The card layer. Its layout changes with data view on <html>. */
    .card-layer { padding:24px; display:grid; gap:16px;
                  grid-template-columns: repeat(auto-fill, minmax(260px,
1fr)); }
    html[data-view="list"] .card-layer { grid-template-columns: 1fr;
max-width:640px; }

    .card { background:var(--panel); border:1px solid var(--line);
            border-radius:14px; padding:18px; cursor:pointer;
            transition: border-color .15s, transform .15s; }
    .card:hover { border-color:var(--accent); transform:
translateY(-2px); }
    .card-eyebrow { font-size:.72rem; text-transform:uppercase; letter-
spacing:.08em;
                    color:var(--accent); margin:0 0 6px; }
    .card-title { font-size:1.1rem; margin:0 0 8px; }
    .card-body { color:var(--muted); margin:0; font-size:.92rem; }

    /* Detail modal */
    .modal-root { position:fixed; inset:0; background:rgba(0,0,0,.6);
                  display:grid; place-items:center; padding:24px; }
    .modal-root[hidden] { display:none; }
    .modal { background:var(--panel); border:1px solid var(--line);
             border-radius:16px; padding:28px; max-width:520px;
width:100%; }
    .modal-close { float:right; background:none; border:none;
color:var(--muted);
                   font-size:1.4rem; cursor:pointer; line-height:1; }
  </style>
</head>
<body>
  <header class="topbar">
    <div class="brand">Pip<span>.</span></div>
    <div class="view-toggle">
      <button data-view-mode="cards">Cards</button>
```

```
      <button data-view-mode="list">List</button>
    </div>
  </header>

  <!-- Empty in the source. bootstrap.js fills this at runtime. -->
  <main class="card-layer" id="cardLayer"></main>

  <!-- Empty modal, shown on card click. -->
  <div class="modal-root" id="modalRoot" hidden>
    <div class="modal">
      <button class="modal-close" id="modalClose" aria-
label="Close">×</button>
      <div id="modalBody"></div>
    </div>
  </div>

  <!-- Load the renderer. It will load the behavior when it's done. -->
  <script src="/bootstrap.js"></script>
</body>
</html>
```

Step 3: render the cards from data

Now the renderer. This is Pip's **bootstrap** step: fetch the data, build the DOM, and *only then* load the behavior. Create it:

`pip/bootstrap.js`

JAVASCRIPT

```javascript
(async function () {
  // 1. Fetch the card data from our own API.
  const cards = await (await fetch('/api/cards')).json();

  // 2. Render it into the (currently empty) card layer.
  const layer = document.getElementById('cardLayer');
  layer.innerHTML = cards.map(card => `
    <article class="card" data-slug="${card.slug}">
      <p class="card-eyebrow">${card.eyebrow}</p>
      <h2 class="card-title">${card.title}</h2>
      <p class="card-body">${card.body}</p>
      <template class="card-detail">${card.detail}</template>
    </article>
  `).join('');

  // 3. The cards exist now, so it's safe to load the behavior.
  const script = document.createElement('script');
  script.src = '/shell.js';
  document.body.appendChild(script);
})();
```

Three numbered steps, and the order is the whole lesson. Fetch, render, *then* load behavior. Each card carries `data-slug="${card.slug}"`, which is its address. The `<template class="card-detail">` holds the modal content without displaying it (a `<template>`'s contents don't render; they wait until code asks for them).

Step 4: add the behavior

Finally, the interactive layer. This runs **after** the cards are in the DOM, so it can find them. Create it:

`pip/shell.js`

JAVASCRIPT

```javascript
// This runs AFTER bootstrap.js has rendered the cards.
(function () {
  const html = document.documentElement;

  // --- View mode -----------------------------------------------------
----
  // The current view lives in ONE place: data-view on <html>. CSS reads
it
  // to lay the cards out; we persist the choice so it survives a
reload.
  function setView(mode) {
    html.dataset.view = mode;
    localStorage.setItem('pip-view', mode);
    // Reflect the active button in the toggle.
    document.querySelectorAll('[data-view-mode]').forEach(b =>
      b.classList.toggle('active', b.dataset.viewMode === mode));
  }

  setView(localStorage.getItem('pip-view') || 'cards');

  document.querySelectorAll('[data-view-mode]').forEach(btn => {
    btn.addEventListener('click', () => setView(btn.dataset.viewMode));
  });

  // --- Card detail modal ---------------------------------------------
----
  const modal = document.getElementById('modalRoot');
  const modalBody = document.getElementById('modalBody');

  function openCard(slug) {
    const card = document.querySelector(`.card[data-slug="${slug}"]`);
    if (!card) return;
    const title = card.querySelector('.card-title').textContent;
    const detail = card.querySelector('.card-detail').innerHTML;
    modalBody.innerHTML = `<h2>${title}</h2>${detail}`;
    modal.hidden = false;
  }

  document.querySelectorAll('.card').forEach(card => {
    card.addEventListener('click', () => openCard(card.dataset.slug));
  });
```

```
  document.getElementById('modalClose').addEventListener('click', () ⇒
{ modal.hidden = true; });
  modal.addEventListener('click', e ⇒ { if (e.target ≡≡ modal)
modal.hidden = true; });

  // Expose openCard so we (and later, the AI) can trigger it by name.
  window.pip = { openCard };
})();
```

Save everything, make sure the server is running, and refresh **http://127.0.0.1:8000**. You should see four cards. Click one and a modal opens with its detail. Click "List" and the cards restack into a single column. Reload, and you're still in List view, because the choice was saved. Click "Cards" to get back to the grid.

You just built a working interface, entirely from data, with no AI anywhere in it.

Why it matters for the AI

Look at three decisions we made, and what each one buys us later:

Every card is addressable. `openCard('inbox')` opens the inbox card. Not "the first card," not "the card at these pixels," but the card *named* `inbox`. When the AI wants to open a card in Chapter 7, it won't simulate a mouse; it'll call `window.pip.openCard('inbox')`, the exact function a click already calls. The human path and the AI path are the same path.

View state lives in one place. "What view am I in?" is `html.dataset.view`. CSS reads it, JavaScript reads it, and soon we'll describe it to the AI in one line. There is no second copy to fall out of sync. When the AI switches you to list view, it'll set that same attribute, and the screen will follow because the screen was reading that attribute all along.

Content comes from the server. The cards aren't in the HTML. They're data. In Chapter 4 that data moves to a database; in Chapter 7 the AI changes what you see by changing a row. None of that requires touching the frontend, because the frontend only ever knew how to *render whatever the data says.*

This is the shape the whole book is aiming at, visible already in a page with no intelligence in it.

NORTH STAR · CORA

Cora builds exactly this way, at scale. Its `.card-layer` is **empty in the HTML source**, filled at runtime by a `bootstrap.js` that fetches `/api/cards` and renders them, then dynamically appends `shell.js`, which wires up dragging, the modal pipeline, the view controller, and eventually the AI's action handler. The order is load-bearing: `shell.js` queries `.card-shell` elements at startup, so it *must* run after the cards are rendered. Cora keeps its screen state on `<html>` too, with `data-view`, `data-theme`, and `data-focus`, the same one-source-of-truth discipline you just applied to `data-view`. Pip is a scale model of Cora's front end, built from the same two ideas: render from data, keep state in one place.

The gotcha: load order will bite you

Here is the mistake almost everyone makes, and it produces a bug that *looks* impossible.

Suppose you decide `bootstrap.js` is overkill and move the behavior straight into `index.html`, in a `<script>` in the `<head>`:

```html
<head>
  ...
  <script src="/shell.js"></script>    <!-- WRONG: runs before cards
exist -->
</head>
```

Now `shell.js` runs while `<main id="cardLayer">` is still empty. Its `document.querySelectorAll('.card')` finds **zero** cards, binds **zero** click handlers, and finishes without error. The page loads, the cards appear (bootstrap renders them a moment later), they *look* clickable, and nothing happens when you click. No error in the console. No crash. Just a dead interface that looks alive.

That's why `bootstrap.js` appends `shell.js` *itself*, as its last step, after rendering. The behavior physically cannot run before the elements it needs exist. The rule to internalize:

> **Rendering creates the elements; behavior binds to them. Behavior must always run after rendering.**

Any time a feature "does nothing and throws no error," suspect load order first. It's the single most common way an AIOS front end breaks, precisely because it fails silently.

A second, quieter gotcha: `bootstrap.js` drops `card.detail` into the page with `innerHTML`, raw HTML and all. That's fine *because the data comes from our own server*. We wrote it, we trust it. The day you let untrusted users write card content, that same line becomes a cross-site-scripting hole. We'll flag exactly where to sanitize when we get there. For now: `innerHTML` is safe only for content you control.

What you built

- A `/api/cards` endpoint serving card data (hand-written today, database-backed next chapter).
- A page whose card layer is **empty in source** and filled at runtime.
- A `bootstrap.js` that fetches, renders, then loads behavior, in that order.
- A `shell.js` that adds a view toggle (state on `<html>`, persisted) and a card-detail modal.
- Every card addressable by slug, and an `openCard(slug)` function that both a human click and (soon) the AI can call.

Exercises

1. **Add a fifth card.** Add a new entry to the `CARDS` list in `server.py`, a "Weather" card, say. Save and refresh. You added a card to the interface *without touching any HTML, CSS, or JavaScript.* Sit with that for a second; it's the whole thesis in one action.
2. **Prove the load-order gotcha.** Temporarily add `<script src="/shell.js"></script>` to the `<head>` of `index.html` (leaving bootstrap in place too). Refresh and try clicking a card, and notice it may misbehave or double-bind. Open DevTools and watch. Then remove it. You've now *seen* the failure the chapter warned about, which is worth more than reading about it.
3. **Add a third view mode.** Give the toggle a "Focus" button (`data-view-mode="focus"`) and a CSS rule `html[data-view="focus"] .card-layer { grid-template-columns: 1fr; max-width: 420px; margin: 0 auto; }`. No JavaScript changes needed, because the toggle already reads `data-view-mode` generically. Notice how little you had to add, because state lives in one place.

Checklist before Chapter 4

- [ ] Four cards render at `http://127.0.0.1:8000`.
- [ ] Clicking a card opens its detail modal; the × and backdrop close it.
- [ ] The view toggle switches layouts, and the choice survives a reload.
- [ ] You understand *why* `shell.js` is loaded by `bootstrap.js` and not from `<head>`.
- [ ] Commit your work: `git add . && git commit -m "Chapter 3: data-driven shell with cards, view toggle, modal"`

The interface is real, but its data is trapped in a Python list that resets every time the server restarts. Chapter 4 fixes that. It gives Pip a database, the AI's writable surface, so that changing what the user sees becomes as simple as changing a row.

CHAPTER FOUR

The Data Layer

> *Pip's cards currently live in a Python list that vanishes every time the server restarts. This chapter moves them into a real database (SQLite, one file, zero setup) and in doing so builds the single most important surface in the whole system: the place the AI will write to.*

In Chapter 3 the frontend learned to render whatever the data says. It never cared *where* the data came from: a Python list, a file, a database, all the same to it. We're about to exploit that indifference. We'll swap the hand-written `CARDS` list for a database, and the frontend won't notice. But the database isn't just tidier storage. Look back at the loop from Chapter 1:

```
YOU → SCREEN STATE → AI → SCREEN STATE → ...
```

"Screen state" has to be something the AI can *write to*. When the AI decides to change what you see, it changes the data, and the screen follows. The database is that writable surface. Everything the AI durably changes about your interface, from Chapter 7 onward, it changes here.

The seam

> **The database is the AI's writable surface. The UI renders from it; the AI writes to it; the user sees the result. It is the hinge of the whole loop.**

That framing has a consequence worth stating up front: the database schema *is part of your AI's capabilities*. A card the AI can't find a row for is a card it can't change. A field that isn't stored is a field the AI can't set. When we design the schema, we're designing what the AI will eventually be able to do.

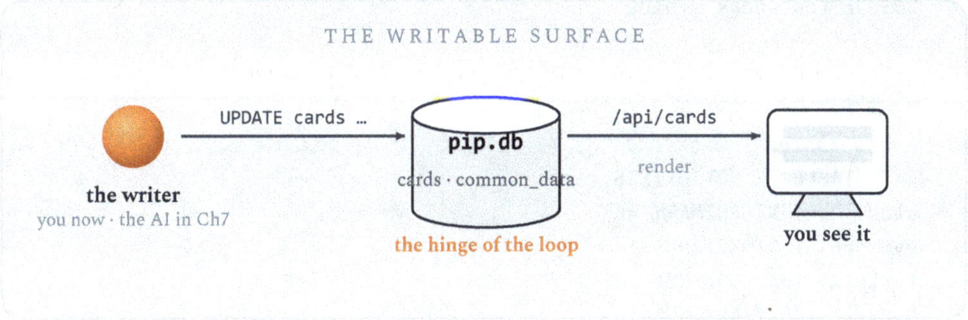

FIGURE 4.1 *The database is the hinge. A write becomes a row, a row becomes what you see. Every durable change runs through here.*

Build the dumb version

We'll use **SQLite**. It's built into Python, so there's nothing to install, and the entire database is a single file, `pip.db`, that we already told git to ignore back in Chapter 2. For a system Pip's size, it's perfect. (Cora uses Postgres; we'll note where that matters.)

Four small pieces: a schema, a connection helper, a one-time setup script, and a rewired endpoint.

Step 1: describe the tables

A schema is just the shape of your data, written in SQL. Create it:

`pip/schema.sql`

```sql
-- One generic lookup table for every "list of allowed values" in Pip.
-- Adding a new category or value is one INSERT, never a schema change.
CREATE TABLE IF NOT EXISTS common_data (
  category TEXT NOT NULL,          -- e.g. 'card_category', 'view_mode'
  value    TEXT NOT NULL,          -- e.g. 'email', 'cards'
  label    TEXT,                   -- human-friendly display name
  UNIQUE (category, value)
);

-- The cards Pip shows. `slug` is the primary key: the id the DOM,
-- this table, and (soon) the AI all use to refer to a card.
CREATE TABLE IF NOT EXISTS cards (
  slug     TEXT PRIMARY KEY,
  eyebrow  TEXT NOT NULL,
  title    TEXT NOT NULL,
  body     TEXT NOT NULL,
  detail   TEXT NOT NULL DEFAULT '',
  category TEXT NOT NULL DEFAULT 'general',
  position INTEGER NOT NULL DEFAULT 0
);
```

Two tables. `cards` is obvious, being the list from Chapter 3 now with a `category` and a `position`. The other one, `common_data`, deserves a real explanation, because it's a pattern you'll reuse constantly.

The `common_data` pattern

Every app accumulates little lists of allowed values: the valid card categories, the valid view modes, later the task statuses and priorities. The naive way is to bake each list into the schema as a `CHECK` constraint (`category TEXT CHECK (category IN ('email','news',...))`). That works until you want to add a value, and then you need a schema migration just to allow one new word.

`common_data` is the alternative. It's a single table holding *every* such list, keyed by `category`:

CATEGORY	VALUE	LABEL
card_category	email	Email
card_category	news	News
view_mode	cards	Cards
view_mode	list	List

Now adding a new card category, or a new view mode, or a whole new *kind* of list, is one `INSERT` : no schema change, no migration, no server edit. The `cards.category` column just stores a plain string; whether it's valid is checked in the application layer against `common_data` . This flexibility is exactly what lets an AIOS grow: as you give the AI new things it can refer to, the vocabulary expands with inserts, not migrations.

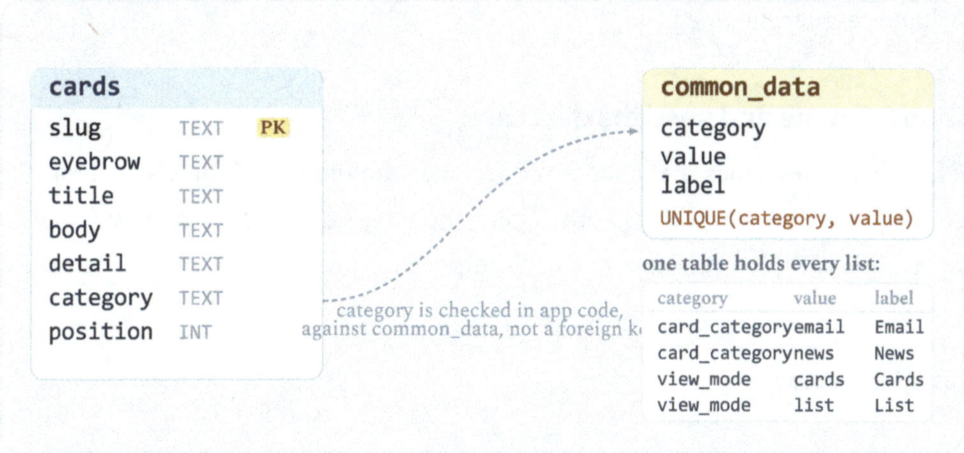

FIGURE 4.2 *cards holds content; common_data holds every "list of allowed values" keyed by category. A new value is one INSERT, never a migration.*

Step 2: a connection helper

You don't want to spell out "open the database, in this folder, with these settings" at every call site. Write it once:

`pip/db.py`

PYTHON

```python
import sqlite3
from pathlib import Path

BASE = Path(__file__).resolve().parent
DB_PATH = BASE / "pip.db"


def get_conn():
    """Open a connection to Pip's SQLite database."""
    conn = sqlite3.connect(DB_PATH)
    # Make rows behave like dicts: row["title"] instead of row[2].
    conn.row_factory = sqlite3.Row
    return conn
```

The one line worth noticing is `conn.row_factory = sqlite3.Row`. Without it, a query returns rows as bare tuples and you're stuck writing `row[2]` and counting columns. With it, you get `row["title"]`, which is readable and safe against column reordering.

Step 3: create and seed the database

The schema describes the tables but doesn't create them, and there's no data yet. A small one-time script does both: it runs the schema, then fills in the starting rows. This is Pip's equivalent of Cora's migration runner, much simpler but the same job.

pip/init_db.py

PYTHON

```python
from db import get_conn, BASE

# The vocabulary lists, as (category, value, label) rows.
SEED_COMMON = [
    ("card_category", "email",    "Email"),
    ("card_category", "news",     "News"),
    ("card_category", "tasks",    "Tasks"),
    ("card_category", "calendar", "Calendar"),
    ("card_category", "general",  "General"),
    ("view_mode",     "cards",    "Cards"),
    ("view_mode",     "list",     "List"),
]

# The starting cards: the same four from Chapter 3, now with a category.
SEED_CARDS = [
    {"slug": "inbox", "eyebrow": "Email", "category": "email",
     "title": "3 threads need a reply",
     "body": "Two from your team, one from a customer asking about
pricing.",
     "detail": "<p>Priya asked for the Q3 numbers. Dana wants to know if
"
               "the Pro plan includes SSO.</p>"},
    {"slug": "briefing", "eyebrow": "News", "category": "news",
     "title": "Your morning briefing is ready",
     "body": "Six stories across AI, markets, and your watchlist.",
     "detail": "<p>Top item: a new open-weights model just shipped.
</p>"},
    {"slug": "tasks", "eyebrow": "Tasks", "category": "tasks",
     "title": "4 things due today",
     "body": "One is overdue. Two can be done in under five minutes.",
     "detail": "<p>Overdue: send the invoice to Dana.</p>"},
    {"slug": "calendar", "eyebrow": "Calendar", "category": "calendar",
     "title": "A clear afternoon",
     "body": "One meeting at 10:00, then open until end of day.",
     "detail": "<p>10:00 standup. After that, nothing scheduled.</p>"},
]


def main():
    conn = get_conn()

    # 1. Create the tables (safe to re-run, thanks to CREATE TABLE IF
NOT EXISTS).
```

```
    conn.executescript((BASE / "schema.sql").read_text())

    # 2. Seed the vocabulary. INSERT OR IGNORE skips rows that already
exist,
    #    so running this script twice does no harm.
    conn.executemany(
        "INSERT OR IGNORE INTO common_data (category, value, label)
VALUES (?, ?, ?)",
        SEED_COMMON,
    )

    # 3. Seed the cards, giving each a position by its order in the
list.
    for position, card in enumerate(SEED_CARDS):
        conn.execute(
            "INSERT OR IGNORE INTO cards "
            "(slug, eyebrow, title, body, detail, category, position) "
            "VALUES (:slug, :eyebrow, :title, :body, :detail, :category,
:position)",
            {**card, "position": position},
        )

    conn.commit()
    conn.close()
    print("Database ready:", (BASE / "pip.db"))


if __name__ == "__main__":
    main()
```

Run it once, from inside `pip/` with your environment active:

```bash
python init_db.py
```

You should see `Database ready: .../pip/pip.db`, and a `pip.db` file will appear in the folder. The whole script is **idempotent**, since every write uses `IF NOT EXISTS` or `INSERT OR IGNORE`, so you can run it again any time without duplicating data or clobbering the tables.

> **A word on `?` and `:name` placeholders.** Notice we never build SQL by pasting values into strings. We pass values *separately*, as `(?, ?, ?)` tuples or `:name` dicts, and SQLite fills them in safely. This is **parameterization**, and it isn't optional politeness. String-pasted SQL is how injection attacks happen. Every query in this book uses placeholders. Make it a reflex.

Step 4: read cards from the database

Now rewire `server.py`. The `/api/cards` endpoint stops returning a hand-written list and starts querying the table. Everything else, the static mount and the health check, stays. And we'll add one small endpoint, `/api/common-data`, so the frontend (and later the AI) can ask what the valid values are.

`pip/server.py`

PYTHON

```python
from pathlib import Path

from fastapi import FastAPI
from fastapi.staticfiles import StaticFiles

from db import get_conn

BASE = Path(__file__).resolve().parent
app = FastAPI(title="Pip")


@app.get("/api/health")
def health():
    return {"status": "ok", "app": "pip"}


@app.get("/api/cards")
def list_cards():
    """Return every card, ordered by position, from the database."""
    conn = get_conn()
    rows = conn.execute(
        "SELECT slug, eyebrow, title, body, detail, category, position "
```

```
            "FROM cards ORDER BY position"
        ).fetchall()
        conn.close()
        return [dict(row) for row in rows]


@app.get("/api/common-data")
def common_data(category: str | None = None):
    """Return lookup values, optionally filtered to one category."""
    conn = get_conn()
    if category:
        rows = conn.execute(
            "SELECT category, value, label FROM common_data "
            "WHERE category = ? ORDER BY value",
            (category,),
        ).fetchall()
    else:
        rows = conn.execute(
            "SELECT category, value, label FROM common_data "
            "ORDER BY category, value"
        ).fetchall()
    conn.close()
    return [dict(row) for row in rows]


# Serve the frontend, after the API routes (so /api/* wins).
app.mount("/", StaticFiles(directory=BASE, html=True), name="static")
```

The `CARDS` Python list is gone entirely. Restart isn't even needed, since `--reload` picks up the change, but the database *does* need to exist, so make sure you ran `init_db.py`. Refresh **http://127.0.0.1:8000**. The four cards render exactly as before.

Nothing on screen changed. Everything underneath did. The frontend code is byte-for-byte the same as Chapter 3; it's still just rendering whatever `/api/cards` returns. But now that data comes from a durable store the AI can write to, and it survives restarts. Stop the server, start it again: the cards are still there.

Why it matters for the AI

Changing a row changes the screen. Open a SQLite shell and edit a card by hand right now, to feel the mechanism the AI will use later:

BASH

```
python -c "import db; c=db.get_conn(); c.execute(\"UPDATE cards SET
title='Edited by hand!' WHERE slug='inbox'\"); c.commit()"
```

On Windows PowerShell the `\"` escaping isn't valid, so use single quotes on the outside and doubled single quotes (`''`) inside instead:

POWERSHELL

```
python -c 'import db; c=db.get_conn(); c.execute("UPDATE cards SET
title=''Edited by hand!'' WHERE slug=''inbox''"); c.commit()'
```

Refresh the page. The inbox card's title changed. You didn't touch HTML, CSS, JavaScript, or the API. You changed one row, and the interface followed. **That is exactly what the AI will do in Chapter 7,** except it'll issue the `UPDATE` in response to you saying "rename my inbox card." The path from AI decision to visible change runs straight through this table.

The vocabulary can grow without a migration. Want a new card category? `INSERT INTO common_data`. Want a new view mode? Same. Because `common_data` holds the lists and the app validates against them, expanding what the AI can refer to never requires touching the schema. An AIOS earns its "OS" by accreting capabilities cheaply, and this is one of the mechanisms that makes that cheap.

The gotcha: decide what the database owns

Here's a decision that quietly shapes the whole system: **not all state belongs in the database.**

Some state is *content*: the cards, and later your tasks and remembered preferences. It's durable, it's the same on every device, and the AI needs to change it. That belongs in `pip.db`.

Other state is *per-device UI fluff*: which view mode you last picked, which panel is collapsed. It's transient, it's specific to the browser you're sitting at, and the AI changing it for you is a momentary action, not a durable fact about the world. That belongs in `localStorage`, which is exactly where Chapter 3 put the view-mode *choice*.

Watch the subtlety in that last one: the *list of valid view modes* lives in `common_data` (it's shared vocabulary the whole system agrees on), but *which one you're currently in* lives in the browser (it's this device, right now). Same feature, two kinds of state, two homes.

Get this line wrong and you feel it immediately. Put transient UI toggles in the database and every trivial click hits the disk and syncs across devices you didn't mean to sync. Put durable content in `localStorage` and the AI can't change it, and it evaporates when the user clears their browser. Draw the line on purpose, per piece of state: *is this a durable fact the AI should be able to change (database), or this-device-right-now UI state (browser)?*

Two smaller gotchas worth banking:

- **Run `init_db.py` before the server, on a fresh clone.** If the cards ever render empty, the first thing to check is whether `pip.db` exists. No database, no cards, and it fails quietly (an empty list is valid JSON).
- **SQLite has one writer at a time.** For a single-user demo that's invisible. But it means SQLite won't carry a busy multi-user app; that's the point where Cora's Postgres earns its keep. Know the ceiling exists so you're not surprised by it later.

NORTH STAR · CORA

Cora's data layer is this same shape, industrialized. It runs **Postgres**, not SQLite: a real database server that handles many concurrent writers, which an AIOS with background agents all writing at once genuinely needs. Its schema changes go through a **versioned migrations runner**, with numbered `.sql` files applied in order and recorded in a `schema_migrations` table, so a schema is reproducible from an empty database and every change is auditable. Your one-shot `init_db.py` is the seed of that idea; the moment Pip's schema changes twice, adopt migrations-as-files. And Cora leans on `common_data` hard: task types, statuses, activity lanes, view modes, card categories, its entire enum vocabulary lives in that one table, validated in the API layer, exactly as you just set up. Pip's data layer is Cora's, scaled down to a single file.

What you built

- A `schema.sql` with a `cards` table and the reusable `common_data` lookup table.
- A `db.py` connection helper with dict-like rows.
- An idempotent `init_db.py` that creates and seeds the database.
- A `/api/cards` endpoint that reads from the database, and a `/api/common-data` endpoint that serves the vocabulary.
- Cards that survive a restart, and a hand-editable row you watched change the screen.
- A clear line between database-owned content and browser-owned UI state.

Exercises

1. **Feel the writable surface.** Using the one-line `python -c` command above as a template, change the `body` of the `briefing` card in the database, then refresh. Then run `python init_db.py` again and notice your change *survives* (because `INSERT OR IGNORE` doesn't overwrite existing rows). You've now edited the interface by editing data, the exact mechanism the AI will use.
2. **Add a category through data alone.** `INSERT` a new row into `common_data` for `('card_category', 'finance', 'Finance')`, then add a card with `category='finance'`. Refresh. You extended Pip's vocabulary *and* its content without changing a single line of schema or code.
3. **Sketch the ownership line.** Write down, for each of these, whether it belongs in the database or in `localStorage`, and why: (a) a card's title, (b) whether the detail modal is currently open, (c) the user's remembered preference "always start in list view," (d) the list of valid view modes. (Answers: a = database/content; b = browser/transient; c = this one's genuinely debatable, since a *durable* preference the AI might set leans database, which is exactly why Chapter 9 puts remembered preferences there; d = database vocabulary, in `common_data`.)

Checklist before Chapter 5

- ☐ `python init_db.py` created a `pip.db` file and printed "Database ready".
- ☐ The four cards still render at `http://127.0.0.1:8000`, now served from the database.
- ☐ You edited a card row by hand and saw the change appear on refresh.
- ☐ `git status` still does **not** list `pip.db` (it was ignored back in Chapter 2).
- ☐ Commit: `git add . && git commit -m "Chapter 4: SQLite data layer with cards + common_data"` (note that `pip.db` won't be committed, which is correct).

Pip now has a face and a memory. What it still lacks is a mind. Chapter 5 wires in the brain: a chat endpoint that isn't a passthrough but a *context-injecting proxy*, the place where, two chapters from now, the AI's eyes and hands will attach.

CHAPTER FIVE

Wire in the Brain

> *This is the chapter where Pip talks. We give it a chat endpoint, a swappable brain (the local or cloud model you set up in Chapter 2), and a place to type. But the important idea isn't "add a chatbot." It's that the chat endpoint is a proxy that injects context, and that's the seam the AI's eyes and hands will attach to in the next two chapters.*

Pip has a face and a memory. Now it gets a mind. By the end of this chapter you'll type "hello" into a box and Pip will answer. That's satisfying, but it's also the least interesting thing this chapter does. The interesting thing is *how* the reply gets made, because we're going to build the chat endpoint not as a dumb pipe from your keyboard to the model, but as a **proxy that wraps every message in context.** Right now that context is just a persona ("you are Pip"). In Chapter 6 it becomes a description of the screen; in Chapter 9 it grows to include Pip's memory of you. The plumbing we lay here is where all of that will flow.

The seam

> **The chat endpoint is not a passthrough. It is a proxy that injects context, the place where perception gets stapled onto every message.**

A naive chat endpoint takes your words and forwards them to the model, verbatim. Ours won't. Every request will pass through the server, which prepends a **system prompt**, the instructions the model reads before it reads you. Today that system prompt is one paragraph. But because every message flows through this one choke point, it's the natural place to inject *anything* we want the model to know each turn. Build the choke point now, well, and the rest of the book just adds to what flows through it.

There's a second seam here, and it's the reason we do a little extra work today: **the brain must be swappable.** You might be on free local Ollama now and a paid cloud model next month, or the reverse. We'll hide that choice behind a single function from the very first line, so switching brains is a one-line change in `.env`, not a rewrite.

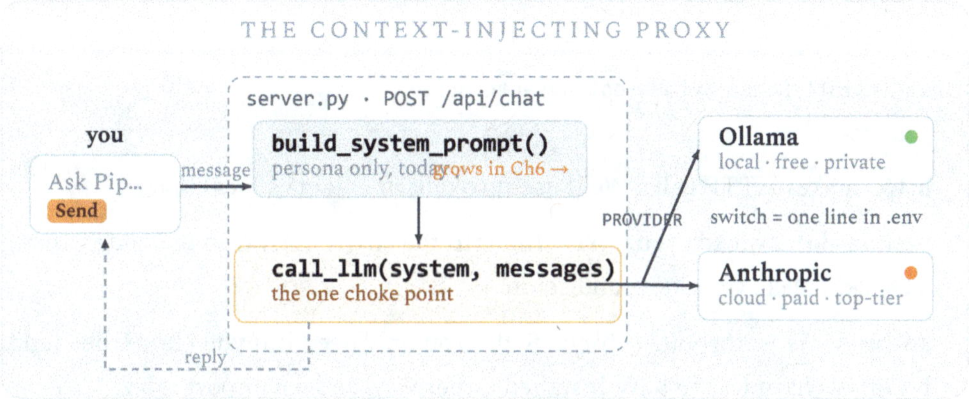

FIGURE 5.1 *Every message flows through one proxy. build_system_prompt() injects context; call_llm() is the single door to a swappable brain.*

Build the dumb version

Three pieces: a provider-dispatching brain module, a chat endpoint that uses it, and a chat box on the page.

Step 1: install the brain libraries

Two new libraries. Add them to `requirements.txt`:

`pip/requirements.txt`

```
fastapi[standard]
uvicorn
httpx
python-dotenv
anthropic
```

Then install:

```bash
pip install httpx python-dotenv anthropic
```

- **httpx** makes HTTP calls. We'll use it to talk to Ollama's local server.
- **python-dotenv** loads your `.env` file into the environment, so the code can read `LLM_PROVIDER`, your keys, and so on.
- **anthropic** is Anthropic's official SDK. You only need it if you chose the cloud brain; it's harmless to have installed either way, and we import it lazily so Ollama users never touch it.

Step 2: the swappable brain

This is the most important file in the chapter. It exposes exactly one function, `call_llm(system, messages)`, and hides the provider choice behind it. Callers never know or care which brain answered.

`pip/llm.py`

PYTHON

```python
import os

import httpx
from dotenv import load_dotenv

# Load .env into the environment so os.environ sees LLM_PROVIDER, keys,
etc.
load_dotenv()

PROVIDER = os.environ.get("LLM_PROVIDER", "ollama")


def call_llm(system: str, messages: list[dict]) -> str:
    """Send a system prompt + conversation to the configured brain,
return its reply.

    `messages` is a list of {"role": "user"|"assistant", "content":
"..."}.
    This is the ONE place the provider choice lives. Callers never
branch on it.
    """
    if PROVIDER == "anthropic":
        return _call_anthropic(system, messages)
    return _call_ollama(system, messages)


def _call_ollama(system: str, messages: list[dict]) -> str:
    base = os.environ.get("OLLAMA_BASE_URL", "http://localhost:11434")
    model = os.environ.get("OLLAMA_MODEL", "llama3.2")
    payload = {
        "model": model,
        "stream": False,
        "options": {"temperature": 0.3},  # low = follows instructions
closely
        # Ollama takes the system prompt as a system-role message.
        "messages": [{"role": "system", "content": system}, *messages],
    }
    # The first call can be slow while Ollama loads the model into
memory.
    resp = httpx.post(f"{base}/api/chat", json=payload, timeout=120.0)
    resp.raise_for_status()
    return resp.json()["message"]["content"]
```

```
def _call_anthropic(system: str, messages: list[dict]) -> str:
    import anthropic  # imported lazily; only needed for the cloud brain

    client = anthropic.Anthropic()  # reads ANTHROPIC_API_KEY from the environment
    model = os.environ.get("ANTHROPIC_MODEL", "claude-opus-5")
    resp = client.messages.create(
        model=model,
        max_tokens=1024,            # room for a full reply without truncating
        system=system,              # Anthropic takes the system prompt separately
        messages=messages,
    )
    # The reply is a list of content blocks; join the text ones.
    return "".join(block.text for block in resp.content if block.type == "text")
```

Study the shape, because it's the pattern from Chapter 1 made real. Both providers accept the same input, a system string and a list of `{role, content}` messages, but they take the system prompt *differently*: Ollama wants it as a message with `role: "system"` at the front of the list; Anthropic wants it as a separate `system=` argument. Each private function adapts. The public `call_llm` doesn't care, and neither does anyone who calls it.

Notice `ANTHROPIC_MODEL` defaults to `claude-opus-5`, the current top-tier model. If you're on the cloud brain and want to spend less on a chatty demo, set `ANTHROPIC_MODEL=claude-haiku-4-5` in your `.env` (Haiku is roughly a fifth the price and plenty for driving a dashboard). That's a choice you make in config; the code just reads it.

If you chose the cloud brain in Chapter 2, add the model line to your `.env` now:

```
ANTHROPIC_MODEL=claude-opus-5
```

Step 3: the chat endpoint

Now the proxy. It builds the system prompt, hands it and the conversation to `call_llm`, and returns the reply. Add this to `server.py`: a Pydantic model to describe the incoming request, a `build_system_prompt()` function (deliberately its own function, because Chapter 6 is going to make it do a lot more), and the endpoint itself.

Add the imports and this code to **`pip/server.py`** (alongside the existing endpoints, before the static mount):

PYTHON

```python
from pydantic import BaseModel

from llm import call_llm


class ChatRequest(BaseModel):
    messages: list[dict]  # [{"role": "user"|"assistant", "content": "..."}]


def build_system_prompt() -> str:
    """Assemble the system prompt sent to the brain on every turn.

    Today it's just a persona. In Chapter 6 this grows to include a live
    description of the screen. This is the injection point.
    """
    return (
        "You are Pip, a warm, concise AI assistant living inside a personal "
        "dashboard. Keep replies short and friendly. When you don't know "
        "something, say so plainly rather than guessing."
    )


@app.post("/api/chat")
def chat(req: ChatRequest):
    reply = call_llm(build_system_prompt(), req.messages)
    return {"reply": reply}
```

Make sure this sits *above* the `app.mount(...)` line. Remember that the static mount is a catch-all, so every real route has to be registered before it.

Step 4: a place to type

Give the page a chat dock. Add this markup to `index.html`, just before the closing `</body>` and the `<script>` tag:

```html
<!-- Chat dock -->
<section class="chat">
  <div class="chat-log" id="chatLog"></div>
  <form class="chat-input" id="chatForm">
    <input id="chatText" placeholder="Ask Pip…" autocomplete="off">
    <button type="submit">Send</button>
  </form>
</section>
```

And add these rules to the `<style>` block in `index.html`:

```css
    .chat { position:fixed; bottom:0; left:0; right:0;
            background:var(--panel); border-top:1px solid var(--line); }
    .chat-log { max-height:40vh; overflow-y:auto; padding:16px 24px;
                display:flex; flex-direction:column; gap:10px; }
    .msg { max-width:70%; padding:10px 14px; border-radius:12px;
           font-size:.92rem; line-height:1.5; white-space:pre-wrap; }
    .msg.user { align-self:flex-end; background:var(--accent);
color:#1a1200; }
    .msg.pip  { align-self:flex-start; background:#1f2634; }
    .chat-input { display:flex; gap:8px; padding:12px 24px;
                  border-top:1px solid var(--line); }
    .chat-input input { flex:1; background:var(--bg); border:1px solid
var(--line);
        color:var(--text); border-radius:10px; padding:10px 14px;
font:inherit; }
    .chat-input button { background:var(--accent); border:none;
color:#1a1200;
        border-radius:10px; padding:0 18px; font:inherit; font-
weight:600; cursor:pointer; }
    /* Keep the cards clear of the docked chat. */
    .card-layer { padding-bottom:200px; }
```

Step 5: wire the chat box to the endpoint

Finally, the behavior. Add this block inside the IIFE in `shell.js` (anywhere after the existing code, before the closing `})();`):

JAVASCRIPT

```javascript
  // --- Chat ---------------------------------------------------------------
---
  const chatForm = document.getElementById('chatForm');
  const chatText = document.getElementById('chatText');
  const chatLog = document.getElementById('chatLog');

  // The running conversation. We send the whole history every turn,
because
  // the model is stateless: it only knows what we tell it each request.
  const history = [];

  function addBubble(role, text) {
    const el = document.createElement('div');
    el.className = 'msg ' + (role === 'user' ? 'user' : 'pip');
    el.textContent = text;
    chatLog.appendChild(el);
    chatLog.scrollTop = chatLog.scrollHeight;
    return el;
  }

  chatForm.addEventListener('submit', async (e) => {
    e.preventDefault();
    const text = chatText.value.trim();
    if (!text) return;
    chatText.value = '';

    addBubble('user', text);
    history.push({ role: 'user', content: text });

    const pending = addBubble('pip', '...');
    try {
      const res = await fetch('/api/chat', {
        method: 'POST',
        headers: { 'Content-Type': 'application/json' },
        body: JSON.stringify({ messages: history }),
      });
      const data = await res.json();
      pending.textContent = data.reply;
```

```
    history.push({ role: 'assistant', content: data.reply });
  } catch (err) {
    pending.textContent = '(Pip could not reach its brain.)';
  }
});
```

Save everything. Make sure your brain is ready, with Ollama running (`ollama run llama3.2` once, to confirm) or your Anthropic key in `.env`. Refresh **http://127.0.0.1:8000**, type "hi, who are you?" into the chat box, and press Send.

Pip answers. You've given it a mind.

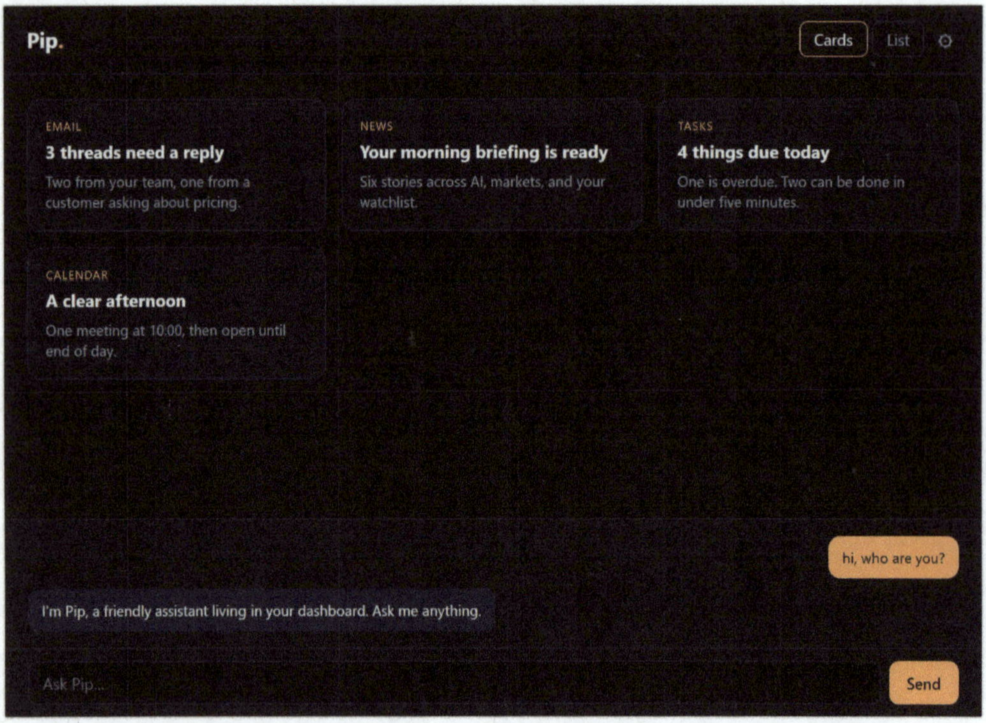

FIGURE 5.2 *Pip's chat dock. You type, Pip answers, and this is the pipe every later feature flows through.*

> **Heads up: the first Ollama reply is slow.** Ollama loads the model into memory on the first request, which can take several seconds (that's why our timeout is a generous 120 seconds). Subsequent replies are fast. If the very first message seems to hang, it's loading, not broken.

The provider abstraction, and why now

You might reasonably ask: I only have one brain configured, so why build the two-provider machinery today? Because the cost of adding it *now* is one `if` statement, and the cost of adding it *later* is hunting down every place that talks to the model and refactoring each one.

Right now there's exactly one caller of `call_llm`. That's the cheapest possible moment to draw the boundary. By Chapter 8, PULSE-style background agents and skills will all want to call a brain too, and because they'll all go through `call_llm`, switching the whole system from local to cloud (or running different agents on different brains) stays a config change. The abstraction isn't speculative complexity; it's the one place where a little structure now prevents a lot of thrash later.

What Pip can and can't do yet

Talk to Pip for a minute and you'll notice something: it's charming, but *blind.* Ask it "what's on my screen?" and it has no idea. It can't see your cards, your view mode, anything. Ask it to "open my inbox card" and it'll cheerfully describe how you might do that, but it can't actually do it.

That's not a bug. That's the state of the system after Chapter 5: Pip has a **mind** but no **eyes** and no **hands**. The system prompt tells it who it is, but nothing about *now*. The next two chapters fix exactly that. Chapter 6 gives it eyes (perception), Chapter 7 gives it hands (control). And both attach right here, at `build_system_prompt()` and `/api/chat`. You've built the seam; now we fill it.

NORTH STAR · CORA

Cora's brain layer is this same `call_llm` idea, expanded into a real dispatch table. Its `agents/llm_clients.py` routes across five providers (`local_ollama`, `local_vllm`, `cloud_anthropic`, `cloud_openai`, `cloud_openrouter`) and stores metadata about each, such as its endpoint and whether it needs a key, so the runtime can resolve where to send a request *and* show a local-vs-cloud badge in the UI. Cora's chat endpoint is a context-injecting proxy exactly like Pip's `/api/chat`, only the context it injects is enormous: the full screen state, the skills list, the user's remembered preferences, and a summary of Cora's own architecture, all assembled fresh every turn and prepended to your message. Pip's one-paragraph `build_system_prompt()` is the seed of that whole apparatus.

The gotcha: verify the model actually got your prompt

Here is a bug that has cost real engineers real days, and it's invisible unless you look for it: **the model silently ignoring your system prompt.**

Cora hit exactly this. One model's chat template quietly *dropped* messages tagged with a certain role, so the model received no system prompt at all, and behaved as if it had never been given one. Nothing errored. The model just wasn't Pip. It answered like a generic assistant, and everyone assumed the prompt was too weak, when in fact the prompt was never delivered.

The defense is a thirty-second test you should run right now. Put a distinctive, checkable instruction in your system prompt, something you'll immediately notice if it's obeyed or ignored. Temporarily change `build_system_prompt()` to end with:

```python
    ) + " Always end every reply with the exact words: ~ Pip 🐦"
```

Reload, send any message. If the reply ends with "~ Pip 🐦", your system prompt is reaching the model. If it doesn't, the prompt is being dropped somewhere and *no amount of rewriting it will help*, because the problem is delivery, not wording. (Revert the change after; it was just a probe.)

Make this a habit: whenever the model behaves as if it can't see instructions you're sure you sent, **verify delivery before you touch the wording.** It's the difference between a five-minute fix and a lost afternoon.

Two smaller settings worth understanding, both already in the code:

- **Temperature.** We set Ollama's to `0.3` (low). Temperature controls randomness; low values make the model follow instructions closely and behave predictably, which is what you want for an assistant that will soon be driving a UI. High values make it creative and wandering, which is wrong for this job. (Anthropic's newest models manage this internally, which is why there's no temperature on that call.)
- **`max_tokens`.** We set `1024` on the Anthropic call, enough room for a full reply. Set it too low and replies get truncated mid-sentence; that becomes a real trap in Chapter 7, when a cut-off reply can also cut off an *action* the AI was trying to emit. Give it headroom.

What you built

- A swappable `llm.py` exposing one `call_llm(system, messages)` function over Ollama and Anthropic.
- A `/api/chat` endpoint that is a **context-injecting proxy**, with `build_system_prompt()` as the injection point.
- A chat dock on the page, with a running conversation sent whole to the (stateless) model each turn.
- A working conversation with Pip: a mind, if not yet eyes or hands.
- The habit of verifying that the system prompt actually reaches the model.

Exercises

1. **Change Pip's personality.** Edit `build_system_prompt()` to make Pip terse and a little sardonic, or formal and precise. Reload, chat, and watch the whole feel change from one paragraph. You're feeling how much leverage the system prompt has, leverage we'll aim at the *screen* next chapter.
2. **Run the delivery test.** Add the "~ Pip 🐦" sign-off probe from the gotcha section, confirm it appears, then remove it. Now you know, first-hand, how to check that your prompt is landing, a skill worth more than any single fix.
3. **Swap brains (if you can).** If you have both Ollama installed and an Anthropic key, flip `LLM_PROVIDER` in `.env` between `ollama` and `anthropic`, restart the server, and chat with each. Same interface, same code path, different brain: that's the abstraction paying off. Notice which one follows instructions more reliably; that difference will matter a lot in Chapter 7.

Checklist before Chapter 6

- ☐ The chat box answers when you send a message.
- ☐ You can state which brain answered (check `LLM_PROVIDER` in `.env`).
- ☐ You ran the "~ Pip 🐦" probe and saw the system prompt reach the model.
- ☐ `build_system_prompt()` is its own function in `server.py` (Chapter 6 needs it that way).
- ☐ Commit: `git add . && git commit -m "Chapter 5: chat endpoint as context-injecting proxy, swappable brain"`

Pip can talk, but it's talking blind. Chapter 6 is the first "OS moment." We assemble a description of what's actually on screen and feed it into that system prompt every turn, so that when you ask "what am I looking at?", Pip can answer, because for the first time it can *see*.

CHAPTER SIX

Perception

> *This is the first OS moment. Until now Pip has been a mind in a jar: it can talk, but it has no idea what's on your screen. This chapter gives it eyes. Every time you send a message, the server assembles a fresh description of the current screen and folds it into the system prompt. Ask Pip "what am I looking at?" and, for the first time, it can answer.*

Everything before this chapter was scaffolding. The face, the database, the chat pipe, all of it existed so that *this* could happen. Look at the loop one more time:

```
SCREEN STATE —perceives—▶ AI
```

That arrow is what we build here. It's the difference between an assistant that *guesses* about your app and one that *knows*. And the guiding principle is blunt:

> **The AI is only as smart as what you show it each turn. A blind model guesses; a sighted one acts.**

The seam

Pip's brain, remember, is stateless. The model doesn't remember your last screen, doesn't secretly watch your browser, doesn't know anything except the exact text we hand it in a single request. So "giving Pip eyes" isn't magic. It's *literally assembling a paragraph that describes the screen and pasting it into the system prompt.* If the paragraph is accurate and current, Pip is sighted. If it's stale or missing, Pip is blind. Perception, in an AIOS, is a text-assembly problem.

There's a nice division of labor hiding in here, and it's worth naming before we write any code:

- **The server knows the content truth.** The cards live in the database; the server reads them directly. It doesn't need to ask the browser what cards exist, because it *owns* that fact.
- **The browser knows the transient UI truth.** Which view mode you're in and whether a card detail is open is state that lives in the browser (Chapter 4's ownership line). The server can't see it, so the browser has to *report* it.

So perception is assembled from two sources: what the server reads from the database, and what the browser tells it about the moment. We'll wire both.

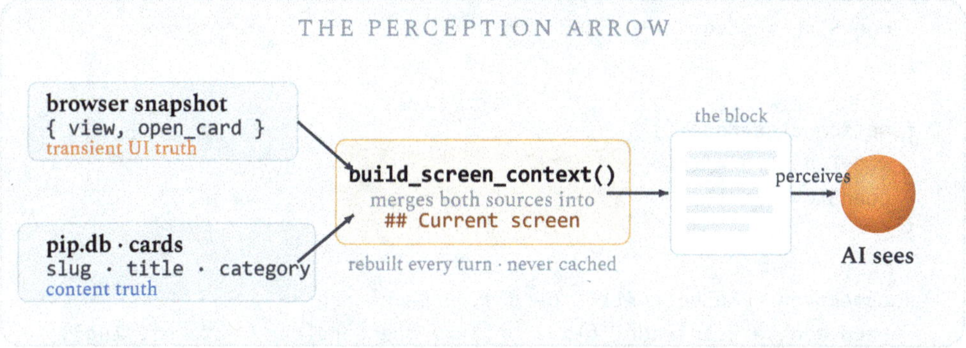

FIGURE 6.1 *Perception is text assembly. Browser state + database content merge into one block, folded into the prompt fresh every turn.*

Build the dumb version

Step 1: let the browser report its state

First, the frontend needs to package up its current UI state and send it along with each chat message. Two small changes to `shell.js`.

First, **track which card is open.** Find the modal section you wrote in Chapter 3 and add a variable that remembers the open card's slug. Update it to look like this:

JAVASCRIPT

```javascript
  // --- Card detail modal ---------------------------------------------
----
  const modal = document.getElementById('modalRoot');
  const modalBody = document.getElementById('modalBody');
  let openSlug = null;    // which card's detail is open, or null

  function openCard(slug) {
    const card = document.querySelector(`.card[data-slug="${slug}"]`);
    if (!card) return;
    const title = card.querySelector('.card-title').textContent;
    const detail = card.querySelector('.card-detail').innerHTML;
    modalBody.innerHTML = `<h2>${title}</h2>${detail}`;
    modal.hidden = false;
    openSlug = slug;                              // remember it
  }

  function closeModal() {
    modal.hidden = true;
    openSlug = null;                              // forget it
  }

  document.querySelectorAll('.card').forEach(card => {
    card.addEventListener('click', () => openCard(card.dataset.slug));
  });
  document.getElementById('modalClose').addEventListener('click',
closeModal);
  modal.addEventListener('click', e => { if (e.target === modal)
closeModal(); });

  window.pip = { openCard };
```

Now add a function that snapshots the current screen state. Put it near the chat code:

JAVASCRIPT

```javascript
// Snapshot the browser's current UI state to send to the server.
function collectScreenContext() {
  return {
    view: document.documentElement.dataset.view || 'cards',
    open_card: openSlug,    // slug of the open card, or null
  };
}
```

Finally, **send that snapshot with every message.** In the chat submit handler, change the `fetch` body to include the screen snapshot:

JAVASCRIPT

```javascript
    const res = await fetch('/api/chat', {
      method: 'POST',
      headers: { 'Content-Type': 'application/json' },
      body: JSON.stringify({
        messages: history,
        screen: collectScreenContext(),    // ← the browser's eyes
      }),
    });
```

That's the browser's half done. It now tells the server, on every turn, what view it's in and whether a card is open.

Step 2: assemble the screen context on the server

Now the server's half. It receives the browser's snapshot, reads the cards from the database, and assembles a single descriptive block. Update the chat code in `server.py`.

First, the request model grows a field:

PYTHON

```python
class ChatRequest(BaseModel):
    messages: list[dict]
    screen: dict | None = None   # the browser's UI snapshot
```

Now the heart of the chapter, a function that builds the screen description:

```python
def build_screen_context(screen: dict | None) → str:
    """Assemble a description of the current screen for the model to
read.

    Content (the cards) comes from the database, the server's own truth.
    Transient UI state (view mode, open card) comes from the browser
snapshot.
    """
    screen = screen or {}

    conn = get_conn()
    cards = conn.execute(
        "SELECT slug, title, category FROM cards ORDER BY position"
    ).fetchall()
    conn.close()

    lines = ["## Current screen"]
    lines.append(f"View mode: {screen.get('view', 'cards')}")
    lines.append(f"Open card detail: {screen.get('open_card') or
'none'}")
    lines.append("Cards currently on screen:")
    for card in cards:
        lines.append(f'- {card["slug"]}: "{card["title"]}" (category:
{card["category"]})')

    return "\n".join(lines)
```

Notice what we send and what we *don't*: slug, title, and category for each card, enough for Pip to reason about them, but **not** the `detail` HTML or the `body` text. Perception should be a lean summary, not a database dump. We'll come back to why in the gotcha.

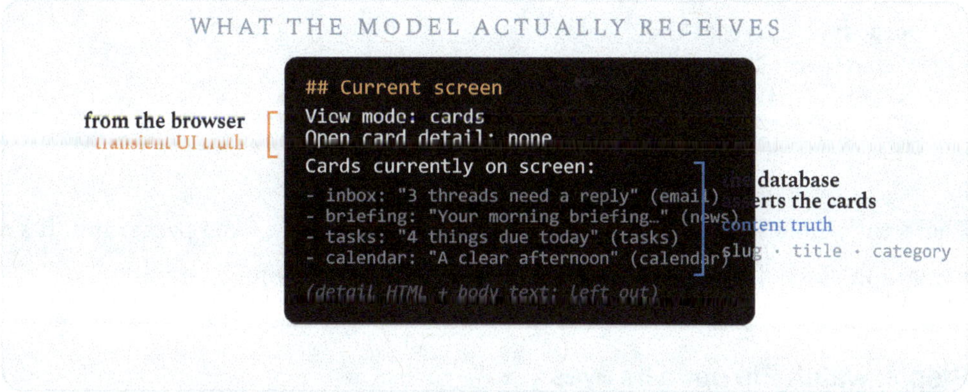

FIGURE 6.2 *The exact block the model reads. Browser lines on top, database lines below, and the heavy fields deliberately left out.*

Now fold it into the system prompt. Update `build_system_prompt` to take the screen and append the context:

```python
def build_system_prompt(screen: dict | None) → str:
    persona = (
        "You are Pip, a warm, concise AI assistant living inside a
personal "
        "dashboard. Keep replies short and friendly. You can SEE the
user's "
        "screen; its current state is described below. Answer questions
about "
        "what's on screen accurately, using the real card names. You
cannot yet "
        "change anything on screen; if asked to, say that's coming
soon."
    )
    return persona + "\n\n" + build_screen_context(screen)
```

And pass the screen through in the endpoint:

```python
@app.post("/api/chat")
def chat(req: ChatRequest):
    reply = call_llm(build_system_prompt(req.screen), req.messages)
    return {"reply": reply}
```

That's the whole change. The system prompt is no longer a fixed paragraph. It's a paragraph *plus a live readout of the screen*, rebuilt on every single request.

Step 3: watch Pip open its eyes

Save everything and refresh. Now have a conversation that would have been impossible one chapter ago:

- Ask: "**What's on my screen right now?**" Pip lists your actual cards by name (inbox, briefing, tasks, calendar) and knows how many there are.
- Ask: "**Which card is about email?**" It answers *inbox*, because it can see the categories.
- Switch to List view, then ask: "**What view am I in?**" It says list.
- Click the tasks card to open its detail, then ask: "**What have I got open?**" It knows the tasks card is open.

None of this is Pip guessing. It's reading the block you assembled, which describes the exact state of the screen at the moment you hit Send. That's perception. The model didn't get smarter; you just started showing it what's in front of it.

Why this is the pivot of the whole book

Stop and appreciate what just changed, because it's the hinge everything else swings on.

Before this chapter, "Pip" was a personality, a way of talking. After this chapter, Pip is *situated.* It exists in a place and knows what's there. That's the precondition for control (next chapter): the AI can't sensibly *change* the screen until it can *see* the screen. "Open my email card" only means something to a model that knows an email card exists and what it's called. You built that knowledge here.

And it's the precondition for everything downstream. Skills, background agents, and memory will each add a section to this same context block. The pattern you just wrote once (read some state, format it into lines, append it to the prompt) is the pattern the rest of the book repeats. Perception isn't one feature; it's the *shape* of every feature that makes Pip aware of something.

NORTH STAR · CORA

Cora's perception is this exact idea, grown to full size. Its `_build_screen_data_block` assembles a rich description **every single turn**: the visible cards, the activity-panel lanes, the available skills, recent skill invocations, the user's remembered preferences, and a live "Currently:" line read straight from the `agent_runs` table showing which background agents are working *right now*. Like Pip's, it's regenerated per message and never cached, so the model always reasons about the present rather than a stale snapshot. Cora even splits the labor the same way you did: durable content comes from Postgres (server truth), transient UI state rides in from the browser. Your `build_screen_context` is a single verse of a long song, but it's the same song.

The gotcha: fresh, but lean

Perception has two failure modes, and they pull in opposite directions.

Failure one: stale context. The whole thing only works because we rebuild the block on *every* request and the browser sends a *fresh* snapshot each time. The tempting "optimization" of building the context once and reusing it, or having the browser send its state only when it changes, quietly breaks perception. The user switches views, sends a message, and Pip describes the *old* view because it's reading a cached block. Rebuild every turn. It's cheap, and correctness depends on it. (This is why `collectScreenContext()` runs at submit time, not once at page load.)

Failure two: bloated context. The opposite temptation is to show the model *everything*: every card's full detail HTML, every field in the database, the entire conversation of your app. Don't. Two reasons. First, cost: on a cloud brain you pay per token, and every token spent describing the screen is a token not spent thinking about the reply, and a token you're billed for. Second, focus: a model buried in irrelevant detail reasons worse, not better, exactly like a person handed a 400-page briefing for a one-line question. That's why Pip's context sends *slug, title, category*, the identifying essentials, and leaves the `detail` bodies out. The model can always be given more when a task actually needs it (a future skill might fetch one card's full detail on demand), but the default readout stays lean.

The rule that reconciles the two:

> **Rebuild the screen context every turn, but include only what the model needs to reason about *now*: what's on screen and, next chapter, what it can do. Not your whole database.**

A third, quieter point: notice that the cards in the context come from the **database**, not from the browser's snapshot. The browser only reports transient UI state (view, open card). This matters, because it means the model can never be fooled into believing in cards that don't exist, since the authoritative list always comes from the server's own source of truth. Let the client report its *own* transient state; let the server assert the *content* truth. Keep that division and your perception stays trustworthy even if the frontend has a bug.

What you built

- A `collectScreenContext()` in the browser that snapshots view mode and the open card, sent with every message.
- A `build_screen_context()` on the server that merges database content (the cards) with the browser's UI snapshot into a fresh, lean description.
- A `build_system_prompt()` that folds that description into the prompt on every turn.
- A Pip that can now *see*, answering "what's on my screen," naming your cards, and knowing your view mode and what's open.
- The instinct to keep perception fresh (rebuilt per turn) and lean (essentials only).

Exercises

1. **Prove it's really seeing.** Add a fifth card to the database (`init_db.py` or a hand `INSERT`), refresh, and ask Pip "how many cards do I have and what are they?" It should report five, by name, because the context reads the live database rather than anything hardcoded. Now you've watched perception track a change you made in the data layer.
2. **Prove freshness matters.** Temporarily change `collectScreenContext()` to always return `{ view: 'cards', open_card: null }` regardless of actual state. Switch to List view, open a card, then ask Pip what view you're in and what's open. Watch it get both wrong, because you fed it a stale snapshot. Revert. You've now *seen* failure-mode one.
3. **Add one field to perception.** Extend the context block with the current time of day, or the total number of cards, or the user's theme (if you added a theme toggle). You add a fact to Pip's awareness by adding one line to `build_screen_context()` , the exact move every future chapter makes.

Checklist before Chapter 7

- ☐ Pip correctly lists your cards by name when asked.
- ☐ Pip correctly reports your current view mode and which card is open.
- ☐ The screen context is rebuilt on every request (not cached).
- ☐ The context sends card summaries (slug/title/category), not full detail bodies.
- ☐ Commit: `git add . && git commit -m "Chapter 6: perception, live screen context injected every turn"`

Pip can see the screen. What it still can't do is *touch* it. Ask it to open a card and it politely declines. Chapter 7 is the second OS moment and the heart of the book: we give Pip hands, through a small, safe vocabulary of actions it's allowed to perform, and wire each one to the very same controls you click.

CHAPTER SEVEN

Control

> *This is the second OS moment and the heart of the book. Pip can see the screen; now we give it hands. By the end of this chapter you'll type "switch to list view" into the chat box and the layout will change, because the AI reached for the same control you click. Read this chapter twice. Everything before it was preparing for this, and everything after it builds on it.*

Giving an AI hands is the most powerful and the most dangerous thing you will build in this whole project. Powerful, because it's the moment Pip stops being a clever narrator and becomes something that *acts*. Dangerous, because "let the AI do things to my computer" is exactly the sentence that should make you nervous. The entire craft of this chapter is getting the power without the danger, and the answer is one idea:

> **The safety comes from a small, closed vocabulary. Never "let the AI run arbitrary code."**

We will not hand Pip a keyboard and let it type commands. We will give it a short, fixed list of *specific actions*, such as open a card or switch a view, each mapped to something you could already do by clicking. The AI can choose among them and combine them, but it cannot invent new ones, and it cannot do anything a user couldn't. Hold that principle and control is safe. Abandon it and you've built a security hole with a friendly face.

The seam

> **Giving the AI hands is the highest-leverage and highest-risk thing you'll build. The safety comes from a small, closed vocabulary.**

Here's the safety property stated precisely, because it's the thing that lets you sleep at night:

Every action the AI can take maps to something a user could already do by clicking. The AI has no power the user doesn't.

Pip's actions will be: open a card, close a card, switch view mode. A user can do all three with a mouse. When the AI "opens the inbox card," it calls the *exact same function* your click calls, `openCard('inbox')` from Chapter 3. It isn't granted a special backchannel to the system; it's handed the same three buttons you have. That's why a closed vocabulary is safe: the worst the AI can do is click your own buttons in an order you didn't ask for, which is annoying at most and never catastrophic. Compare that to "the AI can run any code," where the worst case is unbounded. The whole discipline of this chapter is refusing to cross that line, no matter how tempting the shortcut.

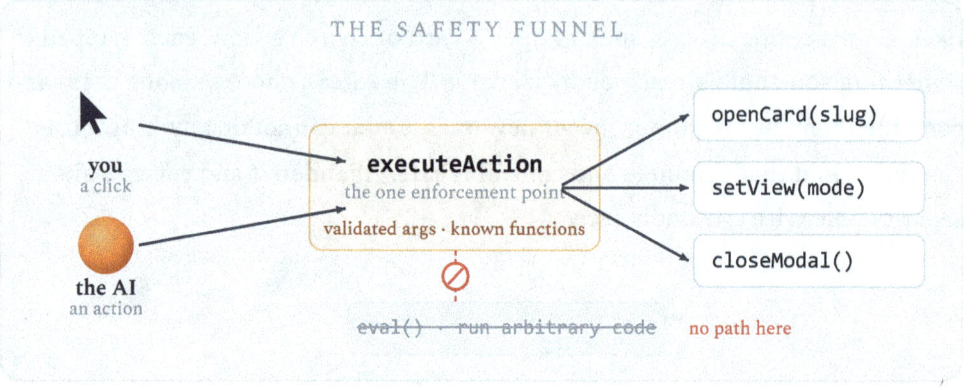

FIGURE 7.B *Human and AI reach the same gate. executeAction runs only known functions with validated args; there is no branch to arbitrary code.*

Build the dumb version

The mechanism has four parts, and we'll build them in order:

1. **Define the vocabulary**, the fixed list of actions.
2. **Teach the model the protocol**, how to request an action.
3. **Build the executor**, the code that carries an action out.
4. **Build the parser**, the code that pulls actions out of the model's reply.

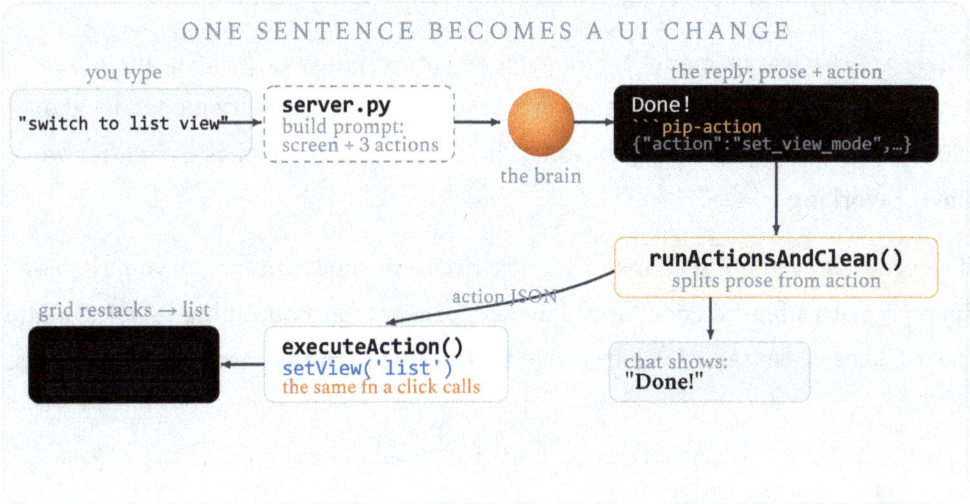

FIGURE 7.A *One sentence, end to end. The reply carries prose plus an action; the browser splits them and runs setView, the very function your click calls.*

Step 1: define the vocabulary

Three actions to start. Each has a name and typed arguments:

```
set_view_mode(mode)    →  mode is "cards" or "list"
open_card(slug)        →  slug names a card that exists
close_card()           →  no arguments
```

That's the whole vocabulary. Closed. The AI will compose these, but it can't add a fourth. Keeping the list this short at first isn't a limitation to apologize for. It's the safety mechanism working. You add actions deliberately, one at a time, each one a thing you've decided the AI is allowed to do.

Step 2: teach the model the protocol

The model needs to know two things: which actions exist, and how to request one. We tell it in the screen context (which already lists the cards, so the AI knows valid slugs).

There are two ways a model can request an action, *native tool-calling* and a *text-fence fallback*. We'll build the fence, because it works with **every** model, local or cloud, with no special configuration. (The two-layer story comes right after we have it working.)

The protocol: when Pip wants to act, it writes a normal sentence to you, then ends its reply with a fenced code block labelled `pip-action` containing JSON. Update `build_screen_context` in `server.py` to describe both the cards *and* the actions:

```python
def build_screen_context(screen: dict | None) -> str:
    screen = screen or {}

    conn = get_conn()
    cards = conn.execute(
        "SELECT slug, title, category FROM cards ORDER BY position"
    ).fetchall()
    conn.close()

    lines = ["## Current screen"]
    lines.append(f"View mode: {screen.get('view', 'cards')}")
    lines.append(f"Open card detail: {screen.get('open_card') or
'none'}")
    lines.append("Cards currently on screen:")
    for card in cards:
        lines.append(f'- {card["slug"]}: "{card["title"]}" (category:
{card["category"]})')

    lines.append("")
```

```
    lines.append("## Actions you can perform")
    lines.append(
        "To perform an action, write a short sentence to the user, then
end "
        "your reply with a fenced code block labelled `pip-action`
containing "
        "JSON. You may include more than one action as a JSON array."
    )
    lines.append("Available actions (use ONLY these, and only slugs
listed above):")
    lines.append('- {"action": "set_view_mode", "args": {"mode": "cards"
| "list"}}')
    lines.append('- {"action": "open_card", "args": {"slug": "<a slug
from above>"}}')
    lines.append('- {"action": "close_card", "args": {}}')
    lines.append("")
    lines.append("Example reply:")
    lines.append("Sure, opening your inbox.")
    lines.append("```pip-action")
    lines.append('{"action": "open_card", "args": {"slug": "inbox"}}')
    lines.append("```")

    return "\n".join(lines)
```

And update the persona in `build_system_prompt`, because Pip can act now, so drop the "you can't change anything" line:

```python
def build_system_prompt(screen: dict | None) → str:
    persona = (
        "You are Pip, a warm, concise AI assistant living inside a
personal "
        "dashboard. Keep replies short and friendly. You can SEE the
user's "
        "screen (described below) and you can DRIVE it using the actions
listed "
        "below. When the user asks you to do something you have an
action for, "
        "do it: write a brief sentence and then the action block. Only
ever use "
        "the actions and card slugs listed; never invent new ones."
    )
```

```
    return persona + "\n\n" + build_screen_context(screen)
```

That's the entire backend change. The server doesn't *execute* actions; it just teaches the model to request them. Execution happens in the browser, because the actions drive the browser's UI.

Step 3: build the executor

Now the frontend. The executor is a `switch` statement with one branch per action, and it's the single place the vocabulary is enforced. An action the switch doesn't recognize does nothing. Add this inside the IIFE in `shell.js` (it needs `setView`, `openCard`, and `closeModal`, which are already defined there):

JAVASCRIPT

```javascript
  // --- Action executor ---------------------------------------------------
  // The ONE place actions are carried out. Anything not matched here is
  // ignored: this switch IS the closed vocabulary, enforced.
  function executeAction(a) {
    console.log('[pip-action] execute:', a);   // log everything (see the gotcha)
    switch (a && a.action) {
      case 'set_view_mode':
        if (a.args && (a.args.mode === 'cards' || a.args.mode === 'list')) {
          setView(a.args.mode);
        }
        break;
      case 'open_card':
        if (a.args && a.args.slug) openCard(a.args.slug);  // openCard already guards bad slugs
        break;
      case 'close_card':
        closeModal();
        break;
      default:
        console.warn('[pip-action] unknown action, ignored:', a);
    }
  }
```

Look at what each branch does. It doesn't `eval` anything. It doesn't build a command. It validates the arguments against a fixed set of possibilities (`mode` must be exactly `'cards'` or `'list'` ; a bad slug is silently ignored by `openCard`) and then calls a function that already existed for the human to use. There is no path from "model output" to "arbitrary behavior." That's the safety property, living in code.

Step 4: build the parser

The model's reply is prose *plus*, sometimes, a fenced action block. The parser's job: find the blocks, run them, and strip them out so the user sees clean prose, never raw JSON. Add this inside the IIFE too:

JAVASCRIPT

````javascript
  // --- Action parser ----------------------------------------------------
  // Pull pip-action blocks out of the reply, run them, return the clean text.
  // Deliberately permissive, because models are sloppy about fences (see the gotcha).
  function runActionsAndClean(reply) {
    let text = reply;

    // Match ```pip-action ... ``` (also tolerate ```json or a bare ``` fence).
    const fenceRe = /```(?:pip-action|json)?\s*([\s\S]*?)```/gi;
    text = text.replace(fenceRe, (whole, body) => {
      if (tryRunActionJson(body)) return '';   // it was actions, so strip it
      return whole;                            // not actions, so leave it alone
    });

    // Also catch a bare {"action": ...} the model forgot to fence.
    const bareRe = /\{\s*"action"\s*:[\s\S]*?\}\s*$/;
    if (bareRe.test(text)) {
      const m = text.match(bareRe);
      if (tryRunActionJson(m[0])) text = text.replace(bareRe, '');
    }

    return text.trim();
````

```
  }

  // Try to parse `body` as one action object or an array of them; run
any found.
  // Returns true if it ran at least one action.
  function tryRunActionJson(body) {
    let parsed;
    try {
      parsed = JSON.parse(body.trim());
    } catch (e) {
      console.warn('[pip-action] could not parse block:', body, e);
      return false;
    }
    const actions = Array.isArray(parsed) ? parsed : [parsed];
    let ran = false;
    for (const a of actions) {
      if (a && a.action) { executeAction(a); ran = true; }
    }
    return ran;
  }
```

This is deliberately forgiving. It accepts the canonical `pip-action` fence, a mislabelled `json` fence, or a bare unfenced object. It handles one action or an array. And when it can't parse something, it *logs and moves on* rather than crashing. The gotcha section explains why every one of those forgiving choices is load-bearing.

Step 5: wire it into the chat

Finally, run the parser on every reply. In the chat submit handler in `shell.js`, change the part that displays the reply:

```javascript
      const data = await res.json();
      const clean = runActionsAndClean(data.reply);    // ⟵ run
actions, strip them
      pending.textContent = clean || '(done)';
      history.push({ role: 'assistant', content: clean });    // store
the clean text
```

Two small but real decisions there. We show `clean`, prose only, so the user never sees the JSON. And we store `clean` in the history, not the raw reply, so the model's own action syntax doesn't pollute the conversation it sees next turn.

Step 6: give Pip its hands

Save everything, refresh, and try the thing this whole book has been building toward:

- Type: "**switch to list view.**" Pip replies something like "Done!" and the cards restack into a column. You didn't touch the toggle. The AI did.
- Type: "**open my tasks card.**" The tasks detail modal opens.
- Type: "**close that and go back to card view.**" Two actions in one reply: the modal closes and the grid returns.
- Type: "**which of my cards is about news, and open it.**" Pip reasons over what it *sees* (perception, Chapter 6) to decide *what to do* (control, this chapter), and opens the briefing card.

That last one is the whole system working at once: the AI perceives the screen, decides on an action, and drives the UI, the full loop from Chapter 1 running in front of you. Open your browser's DevTools console and you'll see `[pip-action] execute:` lines logging exactly what fired. That log is your window into the AI's hands.

The two layers, and why the fence is load-bearing

You just built the **text-fence** layer. There's a second, fancier layer worth understanding, and knowing when each applies.

Layer 1: native tool-calling (preferred, when available). Modern cloud models (and some local ones) support *structured tool use*: you send the model a formal list of "tools" (your actions, as JSON schemas), and the model replies with a clean, structured `tool_use` object instead of text you have to parse. It's more reliable, with no fence to mislabel and no JSON to hand-parse. If you're on the Anthropic brain, you can pass your actions as `tools=[...]` and read structured `tool_use` blocks from the response. It's the better mechanism where the model supports it well.

Layer 2: the text-fence fallback (what you built). Local models, and smaller models generally, often *don't* support tool-calling reliably. They'll return empty tool blocks, or ignore the feature. For those, the fence is the only thing that works: any model that can write text can write a fenced JSON block. That's why we built the fence *first*. It's the universal mechanism, the one that works on the free local brain a beginner is most likely running.

Here's the counterintuitive part, straight from Cora's experience: **in practice, the fence is often the load-bearing path, not the backup.** With a local model, the fence isn't a fallback you rarely hit. It's the mechanism doing all the work, every turn. So build it well, make it robust, and treat native tool-calling as an upgrade you switch on when your model earns it. Pip works today, on any brain, because the fence is solid.

NORTH STAR · CORA

Cora runs both layers of this exact design. It has **13 actions** (`open_agent_tab`, `invoke_skill`, `focus_skill_section`, `open_panel`, `click_card`, `set_view_mode`, `open_settings`, `toggle_panel`, `open_url`, `persona_add`, and more) defined once in Python (a `CORA_TOOLS` schema) and branched once in JavaScript (`executeCoraAction`, the big sibling of your `executeAction` switch). Every chat call sends the tools for native tool-calling *and* the system prompt teaches the fence, so both layers are live. And Cora confirms the lesson above: with its current local-model config, "the text fence is the load-bearing path." Its parser is multi-pass and permissive for the same reason yours is, catching the canonical fence, a mislabelled `json` fence, and bare JSON in prose, and it logs every parse and execute to the console so DevTools shows exactly what fired. Your Pip is a faithful small-scale copy of a mechanism a production AIOS relies on.

The gotcha: models are sloppy, so your parser must be forgiving

Here is where control breaks in the real world, and it's almost never a logic bug. It's that **models produce slightly malformed output, and a strict parser turns "slightly malformed" into "silently did nothing."**

A local model asked to emit a `pip-action` block will, often enough to matter:

- forget the trailing newline before the closing `` ``` ``,
- label the fence `` ```json `` instead of `` ```pip-action ``,
- wrap the whole thing in extra prose,
- emit the JSON with a trailing comma or a smart quote,
- or forget the fence entirely and just write the bare object.

If your parser demands the exact canonical form, every one of those becomes an action that *doesn't fire*, and the user watches Pip say "Done!" while nothing happens. Worse, there's no error: the reply looked fine, the action just vanished. That's the same silent-failure family as the load-order bug in Chapter 3, and it's just as maddening.

The defenses, all of which you built in:

- **Match loosely.** Accept the canonical fence, a `json` fence, and a bare object. Try multiple patterns.
- **Fail soft.** When a block won't parse, log it and move on rather than throwing. One malformed block shouldn't kill the reply.
- **Log everything.** Every `[pip-action]` console line is a breadcrumb. When an action doesn't fire, the log tells you whether the model emitted something you failed to parse, or emitted nothing at all, and those are two very different fixes.

The rule:

> **Make your action parser permissive: match loosely, try multiple patterns, fail soft, and log every attempt. A strict parser makes the AI look broken when it was actually obeying you.**

And two safety reminders that must never slip, no matter how permissive the parsing gets:

- **The executor stays strict even though the parser is loose.** Permissive *parsing* (accept sloppy formatting) is not permissive *execution* (do whatever the JSON says). The `switch` only ever runs known actions with validated arguments. Loose in, strict out.
- **Never `eval`, never build a command from model output.** The instant an action branch does anything other than call a fixed, known function, you've reopened the door the closed vocabulary was built to keep shut. If you ever feel tempted to add an `execute_code` or `run_command` action "just to make Pip more capable," that is the moment to stop and reread the top of this chapter.

What you built

- A closed vocabulary of three actions, described to the model in the screen context.
- A protocol (a `pip-action` fenced block) that any model, local or cloud, can emit.
- An `executeAction` switch, the single enforcement point, that validates arguments and calls the same functions a human click calls.
- A permissive, fail-soft, logged parser that pulls actions from replies and strips them from view.
- A Pip that drives its own UI from chat: the full perceive-decide-act loop, running.
- The discipline of loose parsing, strict execution, and never running arbitrary code.

Exercises

1. **Add a fourth action.** Give Pip a `scroll_to_card(slug)` action: add it to the vocabulary in `build_screen_context`, add a branch in `executeAction` that calls `document.querySelector(...).scrollIntoView()`, and try "take me to my calendar card." You just extended the AI's hands by editing exactly two places, the description and the switch. That two-place symmetry is the pattern.
2. **Break the parser on purpose.** Temporarily make `runActionsAndClean` require the exact `pip-action` label only (drop the `json` and bare-object cases). Then ask Pip to do something a few times until a local model mislabels the fence, and watch the action silently fail while the prose says it worked. Restore the permissive version. You've now *seen* why forgiving parsing is load-bearing.
3. **Find the safety boundary.** Look at your `executeAction` switch and write down, for each branch, the worst thing a malicious or malfunctioning model could cause it to do. You'll find the answer is always "annoy the user by clicking their own buttons," never anything worse. That bounded worst-case *is*

4. the closed-vocabulary guarantee. Now imagine adding an `open_url` action: what new risk appears, and how would you contain it? (This is exactly a question Cora had to answer.)

Checklist before Chapter 8

- ☐ "Switch to list view" changes the layout from chat.
- ☐ "Open my tasks card" opens the modal from chat.
- ☐ A two-action request (close + switch view) performs both.
- ☐ The DevTools console shows `[pip-action]` logs when actions fire.
- ☐ You can explain why the parser is permissive but the executor is strict.
- ☐ Commit: `git add . && git commit -m "Chapter 7: control, a closed action vocabulary the AI drives the UI with"`

Pip can now see and act. The next chapters make it *bigger*. Chapter 8 gives it skills and specialized background agents, capabilities that are both a button you click and an action the AI can invoke, and then Chapter 9, the third OS moment, lets Pip remember you and explain itself.

CHAPTER EIGHT

Skills & Agents

> *Pip can see and act. Now we make it bigger in two ways. First, skills: named capabilities that are both a button you click and an action the AI can invoke, the same work with two triggers. Second, agents: specialized workers with their own identity and an audit trail, whose status indicator lights up while they're busy. And a hard-won piece of advice: build one agent all the way through before you build seven.*

An *action* (Chapter 7) is a verb: open a card, switch a view. It's instantaneous and it drives the UI. A *skill* is bigger, a whole capability with a name, a place in the interface, and real work behind it, such as "summarize my briefing," "draft a reply," or "plan my day." The leap this chapter makes is that a skill has *two* ways to fire: a user can click it, and the AI can invoke it. Same code path, two triggers. Get that right and every skill you add is instantly available to both the human and the AI, for free.

The seam

> **An action is a verb; a skill is a whole capability with a UI. And once you have one agent, you'll want several, each specialized, each with its own brain config.**

We'll build this in two movements: skills first (the capability-with-two-triggers mechanism), then agents (giving those capabilities an owner and a visible heartbeat).

Build the dumb version, part 1: skills

Step 1: tables for skills and runs

Add two tables to `schema.sql`. `skills` describes each capability; `agent_runs` is an audit log with one row per invocation, so we (and the UI) can see what ran and when.

Append to **`pip/schema.sql`**:

```sql
-- Named capabilities. Each belongs to an `agent` and carries the task
-- prompt it runs. A skill is both a button and an AI-invocable action.
CREATE TABLE IF NOT EXISTS skills (
  slug   TEXT PRIMARY KEY,
  name   TEXT NOT NULL,
  agent  TEXT NOT NULL DEFAULT 'pip',
  prompt TEXT NOT NULL
);

-- Audit log: one row per skill invocation, whoever triggered it.
CREATE TABLE IF NOT EXISTS agent_runs (
  id         INTEGER PRIMARY KEY AUTOINCREMENT,
  agent      TEXT NOT NULL,
  skill      TEXT,
  status     TEXT NOT NULL,                   -- queued | running |
succeeded | failed
  result     TEXT,
  created_at TEXT NOT NULL DEFAULT (datetime('now'))
);
```

Step 2: seed some skills

Add a seed list to `init_db.py`, and insert it in `main()`. Each skill names an owning **agent**, and we'll use two, `writer` (writing tasks) and `planner` (planning), and carries the prompt it runs.

Add to **`pip/init_db.py`**:

```python
SEED_SKILLS = [
    ("summarize-briefing", "Summarize my briefing", "writer",
     "Summarize the user's news briefing in two short sentences."),
    ("draft-reply", "Draft an inbox reply", "writer",
     "Draft a short, friendly reply to the customer asking whether the
Pro "
     "plan includes SSO. Keep it under 60 words."),
    ("plan-day", "Plan my day", "planner",
     "Using the user's tasks and calendar, propose a simple plan for the
day "
     "as three short bullet points."),
]
```

Register the two agents as vocabulary by appending these two rows to the existing `SEED_COMMON` list (from Chapter 4), so they're seeded alongside the other lookup values:

```python
    ("agent", "writer", "Writer"),
    ("agent", "planner", "Planner"),
```

Then, inside `main()` , after the card seeding, insert the skills:

```python
    conn.executemany(
        "INSERT OR IGNORE INTO skills (slug, name, agent, prompt) "
        "VALUES (?, ?, ?, ?)",
        SEED_SKILLS,
    )
```

Re-run it (it's idempotent, remember):

```bash
python init_db.py
```

Step 3: endpoints to list and invoke skills

A skill, when invoked, does real work: it calls the brain with the skill's prompt, grounded in the user's card data, and records a run. Two endpoints. Add to `server.py`:

```python
def cards_for_prompt() -> str:
    """A compact dump of card content for skills to work from."""
    conn = get_conn()
    rows = conn.execute(
        "SELECT title, category, body FROM cards ORDER BY position"
    ).fetchall()
    conn.close()
    return "\n".join(f'- [{r["category"]}] {r["title"]}: {r["body"]}'
for r in rows)


@app.get("/api/skills")
def list_skills():
    conn = get_conn()
    rows = conn.execute(
        "SELECT slug, name, agent FROM skills ORDER BY agent, name"
    ).fetchall()
    conn.close()
    return [dict(r) for r in rows]


@app.post("/api/skills/{slug}/invoke")
def invoke_skill(slug: str):
    conn = get_conn()
    skill = conn.execute(
        "SELECT slug, name, agent, prompt FROM skills WHERE slug = ?",
(slug,)
    ).fetchone()
    if skill is None:
        conn.close()
        return {"error": "no such skill"}

    # Record the run as it starts.
    cur = conn.execute(
        "INSERT INTO agent_runs (agent, skill, status) VALUES (?, ?,
'running')",
```

```
        (skill["agent"], slug),
    )
    run_id = cur.lastrowid
    conn.commit()

    # Do the work: the owning agent completes the task, grounded in the
cards.
    system = (
        f"You are {skill['agent'].title()}, a specialist agent inside
Pip. "
        f"Here is the user's current data:\n{cards_for_prompt()}\n\n"
        "Complete the task concisely."
    )
    try:
        result = call_llm(system, [{"role": "user", "content":
skill["prompt"]}])
        conn.execute(
            "UPDATE agent_runs SET status='succeeded', result=? WHERE
id=?",
            (result, run_id),
        )
    except Exception as exc:
        result = f"(The {skill['agent']} agent hit an error: {exc})"
        conn.execute("UPDATE agent_runs SET status='failed' WHERE id=?",
(run_id,))
    conn.commit()
    conn.close()

    return {"run_id": run_id, "agent": skill["agent"], "result": result}
```

Notice the shape: mark a run `running`, do the work, mark it `succeeded` (or `failed`). That audit row is what will let the UI show an agent's heartbeat, and, in Chapter 10, light it up live.

Step 4: render the skills panel

Skills are data, so they render the same way cards do: fetched and built in `bootstrap.js`, then wired in `shell.js`. This reinforces the Chapter 3 rule of render first, bind behavior after.

First give the page a panel. Add to `index.html`, right after the `<header>`:

```html
<aside class="skills-panel" id="skillsPanel"></aside>
```

And these styles to the `<style>` block:

```css
    .skills-panel { position:fixed; top:61px; left:0; bottom:0;
width:220px;
        background:var(--panel); border-right:1px solid var(--line);
        overflow-y:auto; padding:16px 12px 200px; }
    .agent-group { margin-bottom:20px; }
    .agent-badge { display:inline-block; font-size:.7rem; text-
transform:uppercase;
        letter-spacing:.08em; color:var(--muted); border:1px solid var(-
-line);
        border-radius:999px; padding:3px 10px; margin-bottom:8px; }
    .agent-badge.working { color:var(--accent); border-color:var(--
accent);
        animation: pulse 1s ease-in-out infinite; }
    @keyframes pulse { 50% { box-shadow:0 0 0 4px rgba(251,146,60,.18);
} }
    .skill { display:block; width:100%; text-align:left;
background:transparent;
        border:1px solid var(--line); color:var(--text); border-
radius:8px;
        padding:8px 12px; margin:4px 0; font:inherit; font-size:.85rem;
cursor:pointer; }
    .skill:hover { border-color:var(--accent); }
    /* make room for the panel */
    .card-layer { margin-left:220px; }
```

Now render it. Add to `bootstrap.js`, after the cards are rendered and before it loads `shell.js`:

```javascript
// Fetch and render skills, grouped by their owning agent.
const skills = await (await fetch('/api/skills')).json();
const byAgent = {};
skills.forEach(s => (byAgent[s.agent] ||= []).push(s));

document.getElementById('skillsPanel').innerHTML =
  Object.entries(byAgent).map(([agent, list]) => `
    <div class="agent-group">
      <div class="agent-badge" data-agent="${agent}">${agent}</div>
      ${list.map(s => `<button class="skill" data-
slug="${s.slug}">${s.name}</button>`).join('')}
    </div>
  `).join('');
```

Step 5: wire the skills, once, for both triggers

Here's the crux. We write **one** function, `runSkill(slug)`, that invokes a skill and shows its result. A button click calls it. The AI's `invoke_skill` action calls it. Same path.

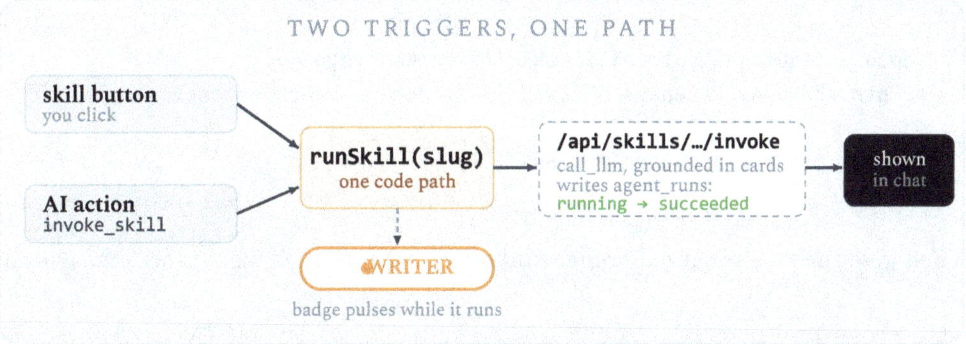

FIGURE 8.A *A click and the AI's action converge on one runSkill(). Same endpoint, same brain call, same audit row, same badge.*

Add to `shell.js`, inside the IIFE:

JAVASCRIPT

```javascript
  // --- Skills ------------------------------------------------------------
  // ONE path to run a skill. A human click and the AI both come through here.
  async function runSkill(slug) {
    const btn = document.querySelector(`.skill[data-slug="${slug}"]`);
    const badge = btn && btn.closest('.agent-group').querySelector('.agent-badge');
    if (badge) badge.classList.add('working');                  // heartbeat on

    const pending = addBubble('pip', `Running "${slug}"…`);
    try {
      const res = await fetch(`/api/skills/${slug}/invoke`, { method: 'POST' });
      const data = await res.json();
      pending.textContent = data.result || '(no result)';
    } catch (err) {
      pending.textContent = '(the skill could not run)';
    } finally {
      if (badge) badge.classList.remove('working');        // heartbeat off
    }
  }

  document.querySelectorAll('.skill').forEach(btn =>
    btn.addEventListener('click', () => runSkill(btn.dataset.slug)));

  window.pip = { openCard, runSkill };
```

Then give the AI the same handle. Add one branch to the `executeAction` switch:

JAVASCRIPT

```javascript
      case 'invoke_skill':
        if (a.args && a.args.slug) runSkill(a.args.slug);
        break;
```

And teach the model that skills exist. In `build_screen_context` (server), add a skills section so the AI knows the valid slugs, the same "grow the context by one section" move you've now made three times:

PYTHON

```
    conn = get_conn()
    skills = conn.execute("SELECT slug, name, agent FROM skills ORDER BY
agent").fetchall()
    conn.close()
    if skills:
        lines.append("")
        lines.append("## Skills you can invoke")
        lines.append('Invoke with: {"action": "invoke_skill", "args":
{"slug": "<slug>"}}')
        for s in skills:
            lines.append(f'- {s["slug"]}: {s["name"]} (agent:
{s["agent"]})')
```

Step 6: see it work, both ways

Save, refresh. You now have a skills panel with two agent groups.

- **Click** "Summarize my briefing." The Writer badge pulses, a `Running...` bubble appears in chat, and a moment later Writer's summary replaces it. You clicked a button; a skill ran; an agent worked.
- **Type** into chat: "**draft me a reply to that customer.**" Pip recognizes the `draft-reply` skill from its context, emits an `invoke_skill` action, and the *exact same thing happens,* with the same badge pulse and the same result, because the AI came through `runSkill`, the same door your click uses.

One capability, two triggers, one code path. That's the whole idea, and you can watch both triggers land in the same place.

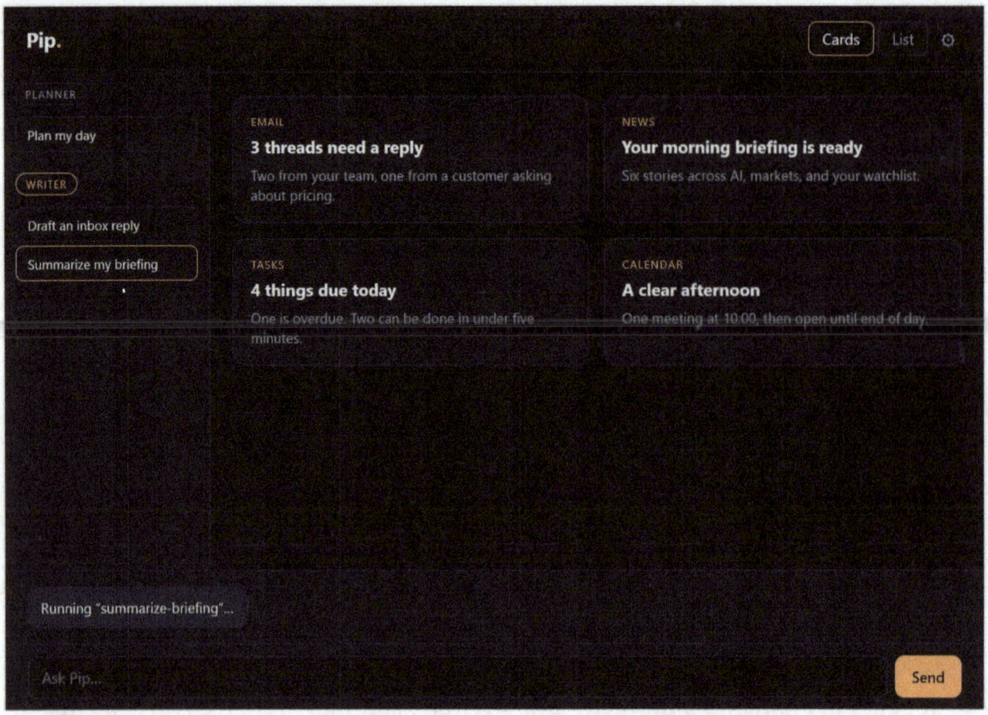

FIGURE 8.B *The skills panel mid-run. Clicking a skill lights its agent's badge while the work is in flight.*

Build the dumb version, part 2: agents with a heartbeat

You've quietly built the agent layer already. Each skill names an owning agent; each invocation writes an `agent_runs` row; the badge pulses while work is in flight. That's an *agent*: a named specialist, with an audit trail, and a visible heartbeat.

Right now the heartbeat is driven by the browser, since the badge pulses while the `fetch` is open. That's honest and it works, but it only knows about work *this browser* started. In Chapter 10 we'll upgrade it: the `agent_runs` table will push changes to the browser over a live channel, so the badge lights up whenever *any* run starts, including background work the user didn't trigger. The audit table you built here is what makes that possible. You wrote the durable record now; Chapter 10 makes it live.

NORTH STAR · CORA

Cora has **seven** agent identities (CORA, ATLAS, SCRIBE, FORGE, PULSE, SIGNAL, CHRONOS) each a tab across the top of the screen with a periodic-table-style two-letter glyph and its own signature colour. Each tab **pulses while that agent has an in-flight `agent_runs` row**, exactly like Pip's badge, but driven live by Cora's push pipeline rather than a local fetch. Skills work the same way too: 14 seeded skills across five categories, each a button that `POST` s `/api/skills/{slug}/invoke` , *and* invocable by the AI through the `invoke_skill` action, one code path for both. Cora's agents each carry their own brain config (an `agent_configs` row holding provider, model, prompt, and temperature), so different agents can run on different models, the natural next step for Pip's simple `agent` label.

The gotcha: build one agent end-to-end before you build seven

This is the most valuable advice in the chapter, and it's about restraint.

When you see Cora's seven glowing agent tabs, the instinct is to scaffold all seven at once: seven names, seven badges, seven stubs. Resist it, hard. **Build one agent that does its whole job (data in, work done, result shown, heartbeat lit) before you add a second.**

Here's why. The first agent is where you discover all the real seams: how work gets triggered, where the run gets recorded, how the result comes back, how the UI reflects it, what happens when it fails. Those seams are the hard part, and you only find them by pushing one agent all the way through. Once the first one works end-to-end, the others are *copies*: a new prompt, a new badge, done. Seven half-built stubs teach you nothing except that you have seven things that don't quite work.

Cora itself follows this: of its seven agents, exactly **one, PULSE, is fully wired**, with its whole pipeline of sources → poll → summarize → digest → schedule, each step wrapped in an `agent_runs` row so its tab glows. The other six are named and visible but not yet complete. That isn't laziness; it's the right order. One complete vertical slice teaches you more than six façades.

For Pip, we built two small agents (Writer and Planner) that genuinely run their skills end-to-end. That's already past the discipline's minimum. When you add a third, you'll notice it takes minutes, because the machinery is done and you're just adding a prompt.

> **Build one agent completely (trigger, work, result, status) before building the next. The first teaches you the seams; the rest are copies.**

What you built

- `skills` and `agent_runs` tables, and seeded skills owned by two agents.
- `/api/skills` and `/api/skills/{slug}/invoke`, where the invoke endpoint records a run, does real brain-backed work, and marks the run's outcome.
- A skills panel rendered from data (bootstrap) and wired once (shell).
- **One** `runSkill(slug)` function that both a click and the AI's `invoke_skill` action call.
- Agent badges that pulse while their agent is working, a heartbeat backed by the `agent_runs` audit log.

Exercises

1. **Add a third skill to an existing agent.** Give Planner a `weekly-review` skill (add it to `SEED_SKILLS`, re-run `init_db.py`). Notice you wrote *only* a data row, with no new endpoint and no new frontend code, and it appears in the panel, works on click, and becomes invocable by the AI, all for free. That's the two-triggers-one-path payoff.
2. **Invoke a skill by voice of the AI.** Ask Pip, in plain English, to do something that matches a skill ("plan out my day for me") and watch it emit `invoke_skill` and light the right agent's badge. Then ask for something *no* skill covers and notice it doesn't hallucinate a skill, because its context lists only the real slugs.
3. **Read the audit trail.** After running a few skills, query the log: `python -c "import db; [print(dict(r)) for r in db.get_conn().execute('SELECT agent, skill, status FROM agent_runs ORDER BY id').fetchall()]"`. You're looking at the record every run left behind, the same table Chapter 10 will turn into a live heartbeat.

Checklist before Chapter 9

- [ ] The skills panel shows two agents with their skills.
- [ ] Clicking a skill pulses its agent's badge and shows a result in chat.
- [ ] Asking Pip to do a skill in plain English runs the *same* skill the same way.
- [ ] `agent_runs` has a row for each invocation (check with the query above).
- [ ] Commit: `git add . && git commit -m "Chapter 8: skills (button + AI action) and agents with run audit + heartbeat"`

Pip can see, act, and now run specialized skills through named agents. The last piece of the three-loop foundation is self-awareness. Chapter 9 lets Pip *remember you*, persisting preferences it applies on every turn, and *explain itself*, answering "how are you built?" from its own architecture instead of inventing an answer.

CHAPTER NINE

Self-Awareness & Memory

> *The third and last OS moment. Pip can see the screen and act on it; now it learns to know itself. Two capabilities: it remembers your preferences, persisting them and applying them on every turn, and it can explain its own architecture, answering "how are you built?" from fact instead of invention. This is the cheap-to-add feature that makes a tool feel like a system.*

There's a moment when an assistant stops feeling like a program and starts feeling like a presence, and it's smaller than you'd expect. It's when you say "always keep your answers short" and it *does*, not just now, but tomorrow, and next week. And it's when you ask "wait, how do you actually work?" and instead of a plausible fabrication, you get the truth about your own system. Both of those are this chapter. Neither is much code. Both change the feel of the thing entirely.

> **An AIOS that can explain and adjust itself stops feeling like a tool and starts feeling like a system. This is cheap to add and disproportionately high-impact.**

The seam

Everything Pip knows each turn, it knows because we assembled it into the system prompt: the persona (Chapter 5), the screen (Chapter 6), the actions and skills (Chapters 7 and 8). Self-awareness is two more sections in that same assembly:

- **Remembered preferences**, a list of things the user told Pip to remember, applied on every reply.
- **Pip's own build**, a description of its architecture, so it can answer questions about itself truthfully.

You've made this move four times now. This is the fifth and the most interesting, because one of these sections Pip can *write to itself.*

Build the dumb version, part 1: remembered preferences

Step 1: a table for notes

Preferences are append-only one-liners. Add a table to `schema.sql` :

```sql
-- Things the user told Pip to remember. Applied on every turn.
CREATE TABLE IF NOT EXISTS persona_notes (
  id         INTEGER PRIMARY KEY AUTOINCREMENT,
  text       TEXT NOT NULL,
  created_at TEXT NOT NULL DEFAULT (datetime('now'))
);
```

Re-run `python init_db.py` to create it (no seed needed, since it starts empty).

Step 2: endpoints, with caps

Four small endpoints: list, add, delete one, clear all. The important detail is the **caps**, a limit on note length and note count. We'll see in the gotcha why these aren't optional. Add to `server.py` :

PYTHON

```python
NOTE_MAX_CHARS = 280
NOTE_MAX_COUNT = 30


class NoteIn(BaseModel):
    text: str


@app.get("/api/persona")
def list_notes():
    conn = get_conn()
    rows = conn.execute("SELECT id, text FROM persona_notes ORDER BY
id").fetchall()
    conn.close()
    return [dict(r) for r in rows]


@app.post("/api/persona")
def add_note(note: NoteIn):
    text = note.text.strip()[:NOTE_MAX_CHARS]        # cap length
    if not text:
        return {"error": "empty note"}
    conn = get_conn()
    count = conn.execute("SELECT COUNT(*) AS n FROM
persona_notes").fetchone()["n"]
    if count >= NOTE_MAX_COUNT:                          # cap count
        conn.close()
        return {"error": "note limit reached"}
    cur = conn.execute("INSERT INTO persona_notes (text) VALUES (?)",
(text,))
    conn.commit()
    nid = cur.lastrowid
    conn.close()
    return {"id": nid, "text": text}


@app.delete("/api/persona/{note_id}")
def delete_note(note_id: int):
    conn = get_conn()
    conn.execute("DELETE FROM persona_notes WHERE id = ?", (note_id,))
    conn.commit()
    conn.close()
    return {"ok": True}
```

```
@app.post("/api/persona/clear")
def clear_notes():
    conn = get_conn()
    conn.execute("DELETE FROM persona_notes")
    conn.commit()
    conn.close()
    return {"ok": True}
```

Step 3: let Pip write to its own prompt

Here's the striking part. Two new actions, `persona_add` and `persona_clear`, let the *AI* add and clear notes. When you say "remember that I prefer British spelling," Pip emits a `persona_add` action, the note is saved, and from the next turn on, Pip reads it back in its own prompt and obeys it. Pip is editing its own instructions.

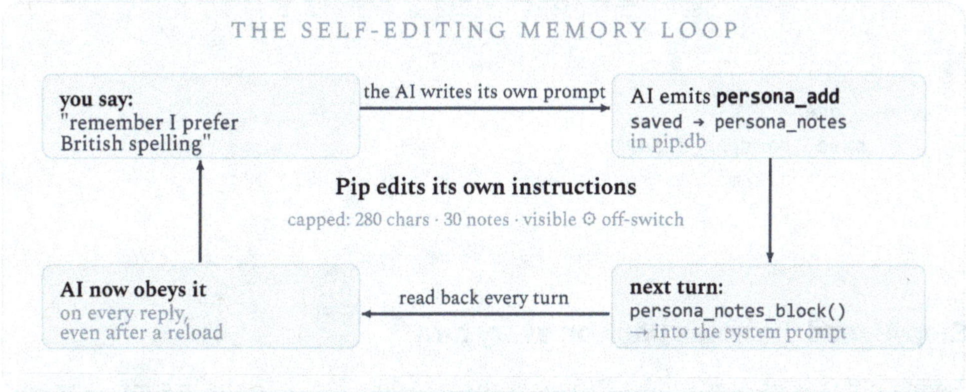

FIGURE 9.B *What Pip saves at your request, it reads back into its own prompt next turn, and obeys. A closed, bounded loop.*

Add the two actions to the vocabulary in `build_screen_context`:

```python
    lines.append('- {"action": "persona_add", "args": {"text": "<one
preference>"}}')
    lines.append('- {"action": "persona_clear", "args": {}}')
    lines.append(
        "Use persona_add ONLY when the user explicitly asks you to
remember a "
        "preference (e.g. 'always be concise', 'call me Captain'). Keep
the "
        "remembered text to a single short instruction."
    )
```

And handle them in the `executeAction` switch in `shell.js`. Unlike the UI actions, these write to the server:

```javascript
      case 'persona_add':
        if (a.args && a.args.text) {
          fetch('/api/persona', {
            method: 'POST',
            headers: { 'Content-Type': 'application/json' },
            body: JSON.stringify({ text: a.args.text }),
          });
        }
        break;
      case 'persona_clear':
        fetch('/api/persona/clear', { method: 'POST' });
        break;
```

Step 4: read the notes back on every turn

Saving notes is useless unless Pip *reads* them. Add a block to the system prompt. In `server.py`:

```python
def persona_notes_block() -> str:
    conn = get_conn()
    rows = conn.execute("SELECT text FROM persona_notes ORDER BY
id").fetchall()
    conn.close()
    if not rows:
        return ""
    lines = ["## Remembered preferences (follow these in every reply)"]
    lines += [f"- {row['text']}" for row in rows]
    return "\n".join(lines)
```

We'll wire it into `build_system_prompt` in a moment, together with part 2.

Build the dumb version, part 2: Pip explains itself

Ask Pip right now "how are you built?" and it will invent something confident and wrong, a generic description of "a typical AI assistant." It has no idea what it actually is. We fix that by handing it a description of its own architecture, every turn.

Step 5: a build summary

Write a short, honest description of how Pip works and make it a constant in `server.py`:

```python
BUILD_SUMMARY = """## About you (Pip's own architecture)
You are Pip, a small AI operating system. How you are built:
- A FastAPI server (server.py) serves a single-page frontend and a JSON
API,
  from one process.
- Cards, skills, remembered preferences, and an agent-run log all live
in a
  SQLite database (pip.db).
- The frontend renders from data: bootstrap.js fetches and renders the
UI,
  then loads shell.js, which holds the behavior.
- You SEE the screen through a context block that is rebuilt every turn,
and
  you DRIVE the UI through a small, fixed set of actions you emit as a
  pip-action code block.
- Skills are capabilities owned by agents (currently Writer and
Planner);
  invoking one records a run in the agent_runs table.
When the user asks how you work or how you were built, answer from these
facts.
Do not invent files, tables, or features that aren't listed here."""
```

That's the whole mechanism. It's just true text about the system, handed to the model so it can answer truthfully. (Cora generates part of this automatically from its changelog, which we'll note shortly, and an exercise adds it to Pip.)

Step 6: assemble it all

Now update `build_system_prompt` to stack every section. This is the full assembly, and it's worth seeing all the pieces the book has added, in order:

PYTHON

```python
def build_system_prompt(screen: dict | None) → str:
    persona = (
        "You are Pip, a warm, concise AI assistant living inside a personal "
        "dashboard. You can see the user's screen and drive it with actions. "
        "Follow the user's remembered preferences below on every reply."
    )
    parts = [
        persona,
        BUILD_SUMMARY,
        persona_notes_block(),          # remembered preferences (may be empty)
        build_screen_context(screen),   # live screen + actions + skills
    ]
    # Drop any empty sections and join.
    return "\n\n".join(p for p in parts if p)
```

Four sections, assembled fresh every turn: who Pip is, what Pip is, what Pip remembers about you, and what's on screen right now. That's a complete, self-aware context.

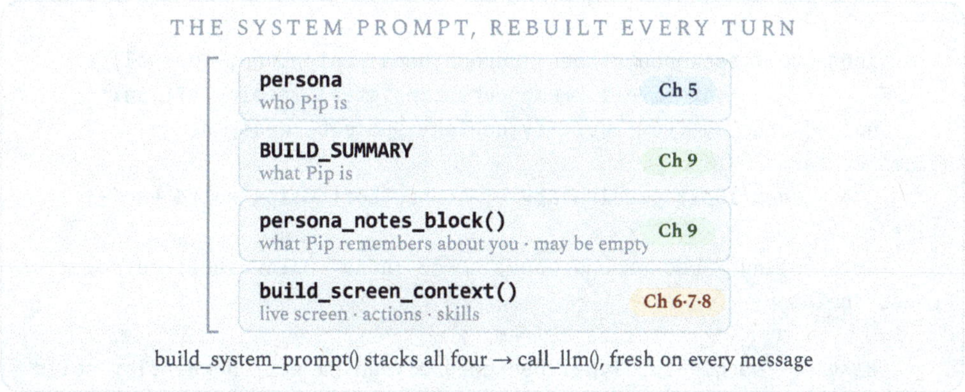

FIGURE 9.A *The whole book, in one prompt. Each chapter added a section; here they stand together, assembled fresh every turn.*

Step 7: a place to review what Pip remembers

Self-modifying memory needs a visible, human-controlled off-switch. Give the settings a small panel. Add a gear button to the header in `index.html`:

```html
<button id="settingsBtn" class="icon-btn" title="What Pip
remembers">⚙</button>
```

A settings modal (near the other modal):

```html
<div class="modal-root" id="settingsRoot" hidden>
  <div class="modal">
    <button class="modal-close" id="settingsClose" aria-
label="Close">×</button>
    <h2>Pip remembers</h2>
    <div id="notesList"></div>
    <button id="notesClear" class="danger">Clear all</button>
  </div>
</div>
```

A little CSS:

```css
.icon-btn { background:none; border:none; color:var(--muted);
            font-size:1.2rem; cursor:pointer; margin-left:8px; }
.note { display:flex; justify-content:space-between; align-
items:center;
        gap:10px; padding:8px 0; border-bottom:1px solid var(--
line); }
.note button { background:none; border:none; color:var(--muted);
cursor:pointer;
               font-size:1.1rem; }
.danger { margin-top:14px; background:transparent; border:1px solid
#b3383c;
          color:#e5484d; border-radius:8px; padding:6px 14px;
cursor:pointer; font:inherit; }
.muted { color:var(--muted); }
```

And the behavior in `shell.js`:

```javascript
  // --- Settings: review what Pip remembers ---------------------------
---
  const settingsRoot = document.getElementById('settingsRoot');

  async function renderNotes() {
    const notes = await (await fetch('/api/persona')).json();
    const list = document.getElementById('notesList');
    list.innerHTML = notes.length
      ? notes.map(n => `<div class="note"><span>${n.text}</span>
          <button data-id="${n.id}" aria-label="Forget">×</button>
</div>`).join('')
      : '<p class="muted">Nothing remembered yet. Tell Pip to remember
something.</p>';
    list.querySelectorAll('button[data-id]').forEach(b =>
      b.addEventListener('click', async () => {
        await fetch('/api/persona/' + b.dataset.id, { method: 'DELETE'
});
        renderNotes();
      }));
  }

  document.getElementById('settingsBtn').addEventListener('click', () =>
{
    settingsRoot.hidden = false;
    renderNotes();
  });
  document.getElementById('settingsClose').addEventListener('click', ()
=> settingsRoot.hidden = true);
  settingsRoot.addEventListener('click', e => { if (e.target ===
settingsRoot) settingsRoot.hidden = true; });
  document.getElementById('notesClear').addEventListener('click', async
() => {
    await fetch('/api/persona/clear', { method: 'POST' });
    renderNotes();
  });
```

Step 8: watch Pip remember and reflect

Save, refresh, and try both halves:

- Tell Pip: "**From now on, always call me Captain.**" It replies "Aye aye!" (or similar) and emits a `persona_add`. Open the ⚙ settings and your note is there. Now send any other message: Pip calls you Captain. **Reload the page and do it again, and it still calls you Captain**, because the preference is in the database, applied on every turn. Open settings and click × to forget it; it stops.
- Ask: "**How do you actually work? What database do you use?**" Pip answers with the truth (SQLite, `pip.db`, the bootstrap/shell split, the `pip-action` protocol) because it's reading `BUILD_SUMMARY`, not guessing. Ask it something *not* in the summary and notice it says it doesn't know, rather than inventing.

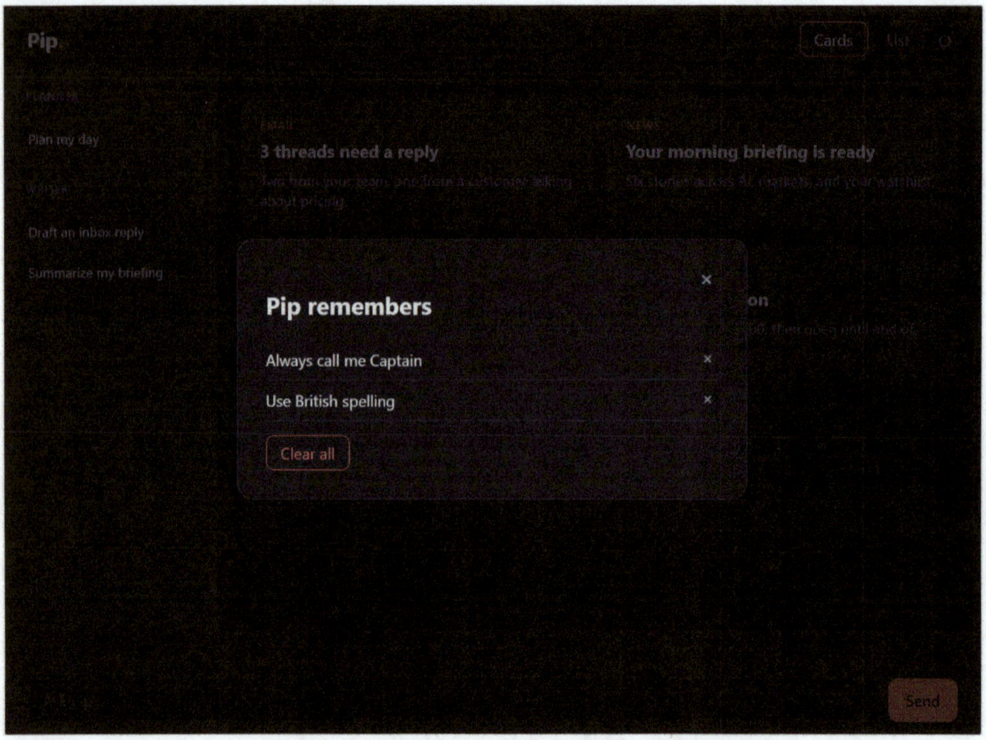

FIGURE 9.C *The Pip remembers panel: every note is visible, removable one by one, and clearable. The off-switch the gotcha insists on.*

Pip now adjusts itself to you and can account for itself. That's the third loop closed.

NORTH STAR · CORA

Cora's version of both halves is the same idea at scale. Its **persona notes** persist in an `agent_settings` row keyed to Cora, capped at 400 characters per note and 50 notes total, added via the exact `persona_add` cora-action you just built, listed in the chat context as a "User-customised preferences" block every turn, and managed through a settings UI with a × per row and a Clear-all. Its **build self-awareness** is two blocks: a hand-written architecture overview (`_BUILD_SUMMARY_BLOCK` , about 1.8K tokens, the big sibling of your `BUILD_SUMMARY`) *and* a `## Recent changes` block that `_recent_changelog_entries` auto-parses from `CHANGELOG.md` , taking the last several entries with title, date, and a one-line summary. So Cora can answer not just "how are you built" but "*when* did we add X," citing the real changelog title. The noted next step is retrieval (RAG) over the full codebase, for questions deeper than a summary can hold. Pip's two constants are the seed of Cora's whole self-knowledge apparatus.

The gotcha: cap everything, and keep the off-switch visible

Self-modifying prompts are genuinely dangerous in one specific, boring way: **they can grow without bound and eat your context.**

Think it through. Every remembered note goes into the system prompt, every turn, forever. If there's no cap, an over-eager model (or a user who says "remember this" a lot) accumulates notes until the "remembered preferences" section is longer than the actual conversation. On a cloud brain you're now paying to re-send a growing pile of preferences on every message; on any brain, the model's attention is increasingly spent on old notes instead of the current question. Left unchecked, an AIOS that remembers *everything* slowly forgets how to *respond,* because its own memory crowds out the moment.

The defenses, which you built in:

- **Hard caps on length and count.** 280 characters per note, 30 notes, enforced server-side. The AI cannot exceed them even if it tries. (Cora uses 400 and 50; pick numbers that fit your token budget.)
- **Append-only, one instruction per note.** Notes are short, discrete preferences, not essays. This keeps each one cheap and the set scannable.
- **A visible off-switch.** The settings panel with per-note delete and Clear-all means the human always has final control over what the AI remembers about them. Self-modification the user can't inspect or undo is a trap; self-modification with a visible list and a delete button is a feature.

> **Cap self-modifying memory hard, in length, count, and a visible clear, or the AI's own preference list grows until it crowds out the conversation.**

One more discipline, in the prompt rather than the code: we told the model to use `persona_add` **only when the user explicitly asks to remember something.** Without that instruction, models tend to "helpfully" remember everything (your name, your mood, that one thing you mentioned once) and the list fills with noise. Remembering should be a deliberate act the user requests, not a reflex the AI indulges. And there's a quiet safety angle: because notes are instructions the AI will obey every turn, you only ever want *the user* writing them. In Pip that's guaranteed, since only you type. In a system that ingested untrusted content, you'd need to be sure that content could never slip a `persona_add` into Pip's memory, a preview of the injection concerns any AIOS eventually faces.

What you built

- A `persona_notes` table and four endpoints (list / add / delete / clear), with hard caps
- Two actions, `persona_add` and `persona_clear`, that let the AI write to its own instructions on your explicit request.
- A `persona_notes_block()` that reads those notes into the prompt every turn, so Pip *applies* what it remembers.
- A `BUILD_SUMMARY` that lets Pip answer questions about its own architecture truthfully.
- A full `build_system_prompt` that stacks persona, build, memory, and live screen into a complete, self-aware context.
- A settings panel to review, delete, and clear what Pip remembers.

Exercises

1. **Prove persistence.** Tell Pip to remember two preferences ("be very brief" and "use British spelling"). Reload the whole page. Chat again, and both still apply, and both appear in ⚙ settings. You've watched the AI's self-edits survive a restart, because they live in the database, not the conversation.
2. **Add a changelog block.** Create a `CHANGES.md` file with a few dated one-line entries, write a small function that reads its last five lines, and append them to the prompt as a `## Recent changes` block (mirroring `persona_notes_block`). Now ask Pip "what did you add recently?" and it cites your changelog. You've just built Cora's `_recent_changelog_entries` in miniature.
3. **Stress the cap.** Temporarily lower `NOTE_MAX_COUNT` to 2, then ask Pip to remember three separate things. Watch the third `persona_add` get rejected by the server, and confirm the settings panel shows only two. You've seen the runaway-prevention working, the exact thing that keeps self-modification from eating the context.

Checklist before Chapter 10

- ☐ Telling Pip to remember a preference persists it (survives reload) and applies it every turn.
- ☐ The ⚙ settings panel lists notes, deletes one, and clears all.
- ☐ Asking "how are you built?" gets a truthful, specific answer.
- ☐ Notes are capped in length and count, server-side.
- ☐ Commit: `git add . && git commit -m "Chapter 9: self-awareness, persona notes + build summary in every prompt"`

You've now built all three loops: Pip can **see** the screen (Chapter 6), **drive** it (Chapter 7), and **know itself** (this chapter). The foundation is complete. The remaining chapters make Pip feel *alive*. Chapter 10 pushes changes to the screen without a reload, so when a background agent finishes, the interface just updates on its own.

CHAPTER TEN

Live Updates

> *The foundation is done, and Pip can see, act, and know itself. The rest of the book makes it feel alive, and this chapter is the biggest step. Right now the screen only changes when you do something. By the end of this chapter, the screen changes on its own: when a background agent finishes, or a card is edited anywhere, every open browser updates itself, with no reload and no refresh button. Push, not poll.*

There's a quiet tell that separates a real system from a demo: a demo waits for you; a system moves on its own. When an agent finishes work in the background, a demo makes you hit refresh to see the result. A system just shows it. That aliveness is what this chapter adds, and the mechanism is simpler than it feels.

> **An AIOS should update without a reload. When a background agent finds something, the screen should just change. Push, not poll.**

"Poll" would mean the browser asking the server "anything new?" every few seconds, which is wasteful and laggy. "Push" means the server *tells* the browser the moment something changes. We'll build the push channel with **Server-Sent Events (SSE)**, the simplest one-way server-to-browser stream, built into every browser, with no extra libraries.

The seam

Everything durable in Pip lives in the database or gets recorded there: the cards, and the `agent_runs` audit log from Chapter 8. The push channel connects *changes to that state* to *the screen*. When something writes a change, it announces it; every connected browser hears the announcement and updates the affected piece of the page in place. Two things to build: a way for the server to broadcast, and a way for the browser to listen and surgically apply.

Recall the loose end from Chapter 8: the agent badge pulsed only in the tab that *started* the work, because the pulse was driven by that tab's own `fetch`. After this chapter, the badge is driven by the server, so it lights up in *every* open tab whenever *any* run starts, including work no browser initiated. That upgrade is the whole point.

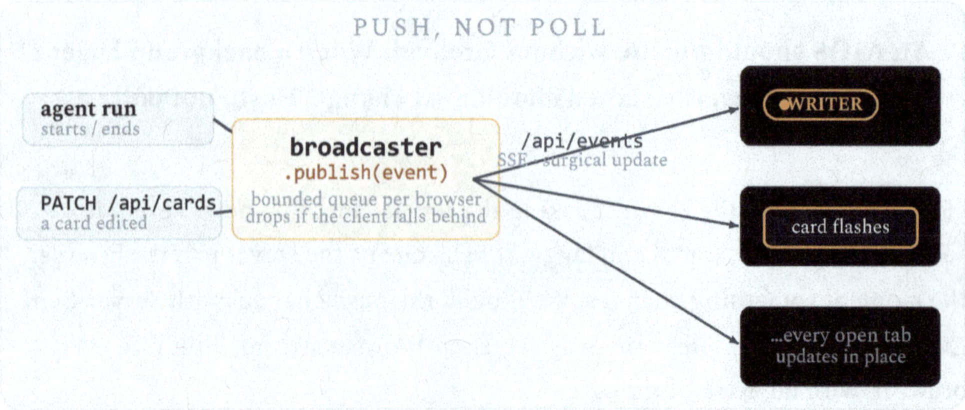

FIGURE 10.1 *One change is announced once and fanned to every screen over SSE. Each tab updates only the affected element, with no reload.*

Build the dumb version

Step 1: a broadcaster on the server

The broadcaster holds one queue per connected browser and fans every event out to all of them. Add this to `server.py` (near the top, after the imports):

PYTHON

```python
import asyncio
import json

from fastapi.responses import StreamingResponse


class Broadcaster:
    """Fans out change events to every connected browser."""

    def __init__(self):
        self.subscribers: set[asyncio.Queue] = set()
        self.loop: asyncio.AbstractEventLoop | None = None

    def subscribe(self) -> asyncio.Queue:
        q = asyncio.Queue(maxsize=100)   # bounded: a slow client can't leak memory
        self.subscribers.add(q)
        return q

    def unsubscribe(self, q: asyncio.Queue):
        self.subscribers.discard(q)

    def publish(self, event: dict):
        """Announce a change. Safe to call from any thread."""
        if self.loop is None:
            return  # no one is listening yet
        for q in list(self.subscribers):
            self.loop.call_soon_threadsafe(self._put, q, event)

    @staticmethod
    def _put(q: asyncio.Queue, event: dict):
        try:
            q.put_nowait(event)
        except asyncio.QueueFull:
```

```
                pass    # drop it; every update is idempotent, so the next
one recovers


broadcaster = Broadcaster()
```

Two design choices carry real weight here, and the gotcha section returns to both. The queue is **bounded** (`maxsize=100`) and full queues **drop rather than block**, so a browser that falls behind can't wedge the server or eat unbounded memory. And `publish` goes through `loop.call_soon_threadsafe` , which is the correct way to hand an event from a *worker thread* (where our skill code runs) to the *async event loop* (where the SSE streams live). More on why in a moment.

Step 2: the SSE endpoint

Now the endpoint browsers connect to. It subscribes a queue and streams each event as it arrives, with a heartbeat every 15 seconds so proxies don't kill an idle connection. Add to `server.py` :

```python
@app.get("/api/events")
async def events():
    # Capture the running loop the first time someone connects;
publish() needs it.
    broadcaster.loop = asyncio.get_running_loop()
    q = broadcaster.subscribe()

    async def stream():
        yield "retry: 3000\n\n"    # tell the browser to reconnect after
3s if dropped
        try:
            while True:
                try:
                    event = await asyncio.wait_for(q.get(),
timeout=15.0)
                    yield f"event: change\ndata:
{json.dumps(event)}\n\n"
                except asyncio.TimeoutError:
```

```
                    yield ": ping\n\n"    # heartbeat keeps the
connection alive
        finally:
            broadcaster.unsubscribe(q)

    return StreamingResponse(stream(), media_type="text/event-stream")
```

The wire format is plain text: `event: change`, then `data: <json>`, then a blank line. That's the entire SSE protocol. A line starting with `:` is a comment (our heartbeat). `retry: 3000` tells the browser's built-in `EventSource` to reconnect automatically after 3 seconds if the stream drops, so reconnection is free.

Step 3: announce changes

An event channel is useless until something publishes. Wire up the two changes worth broadcasting.

First, **agent runs**. In `invoke_skill` (Chapter 8), announce the run starting and finishing. Add the two `publish` calls:

```python
    run_id = cur.lastrowid
    conn.commit()
    broadcaster.publish({"type": "agent_run", "agent": skill["agent"],
"status": "running"})

    # ... do the work ...

    conn.commit()
    conn.close()
    broadcaster.publish({"type": "agent_run", "agent": skill["agent"],
"status": "done"})
    return {"run_id": run_id, "agent": skill["agent"], "result": result}
```

Second, **card edits**. Add an endpoint to update a card, which announces the change so the screen can follow. This is also the endpoint the AI will eventually use to change a card on your behalf, the writable surface from Chapter 4 now made live:

```python
class CardUpdate(BaseModel):
    title: str | None = None
    body: str | None = None


@app.patch("/api/cards/{slug}")
def update_card(slug: str, patch: CardUpdate):
    fields = {k: v for k, v in patch.model_dump().items() if v is not
None}
    if fields:
        conn = get_conn()
        sets = ", ".join(f"{key} = :{key}" for key in fields)   # keys
are fixed model fields
        conn.execute(f"UPDATE cards SET {sets} WHERE slug = :slug",
{**fields, "slug": slug})
        conn.commit()
        conn.close()
        broadcaster.publish({"type": "card", "slug": slug})
    return {"ok": True}
```

Step 4: listen and apply in the browser

The browser side lives in `bootstrap.js`, alongside the other data plumbing. Open an `EventSource`, and route each event to a surgical update, changing *only the affected element* and never reloading the page. Add to the end of the `bootstrap.js` IIFE (after it loads `shell.js`):

```javascript
  // --- Live updates: listen for pushed changes and apply them
surgically -
  const es = new EventSource('/api/events');

  es.addEventListener('change', async (e) => {
    const ev = JSON.parse(e.data);

    if (ev.type === 'agent_run') {
      // Light or clear the agent's badge, in EVERY open tab.
      const badge = document.querySelector(`.agent-badge[data-
agent="${ev.agent}"]`);
      if (badge) badge.classList.toggle('working', ev.status ===
'running');
```

```
    } else if (ev.type === 'card') {
      // Re-fetch and update just this one card, in place.
      const cards = await (await fetch('/api/cards')).json();
      const data = cards.find(c => c.slug === ev.slug);
      const el = document.querySelector(`.card[data-
slug="${ev.slug}"]`);
      if (data && el) {
        el.querySelector('.card-title').textContent = data.title;
        el.querySelector('.card-body').textContent = data.body;
        el.classList.add('flash');                    // a visible
signal
        setTimeout(() => el.classList.remove('flash'), 1200);
      }
    }
  });
```

And a flash style so a live change is *seen*, in `index.html`:

```css
    .card.flash { animation: flash 1.2s ease; }
    @keyframes flash {
      0%   { box-shadow: 0 0 0 3px var(--accent); }
      100% { box-shadow: 0 0 0 0 transparent; }
    }
```

Step 5: watch the screen move on its own

Save, and set up the demo that shows push clearly: **open the page in two browser tabs, side by side.**

- In tab A, click a skill, say "Summarize my briefing." Watch tab **B:** Writer's badge lights up while the skill runs, then goes dark when it finishes. Tab B didn't start anything. The server *told* it. That's push, across clients, with no reload.
- Now edit a card from the terminal, as if a background job had:

```bash
curl -X PATCH http://127.0.0.1:8000/api/cards/inbox \
  -H "Content-Type: application/json" \
  -d '{"title": "Updated live, no reload!"}'
```

  On Windows PowerShell, `curl` is an alias for `Invoke-WebRequest` and the `\` line-continuation isn't valid, so use this instead:

```powershell
Invoke-WebRequest -Uri http://127.0.0.1:8000/api/cards/inbox -Method
PATCH -ContentType 'application/json' -Body '{"title": "Updated live
- no reload!"}'
```

  Both tabs update the inbox card's title *and flash it*, instantly. Nobody refreshed. The change happened in the database, was announced, and every open screen followed.

That second demo is the whole thesis of the chapter in one command: a change to the state became a change on every screen, automatically. When Pip grows real background agents, such as a briefing that refreshes overnight or a monitor that watches for something, this is how their results reach you without you asking.

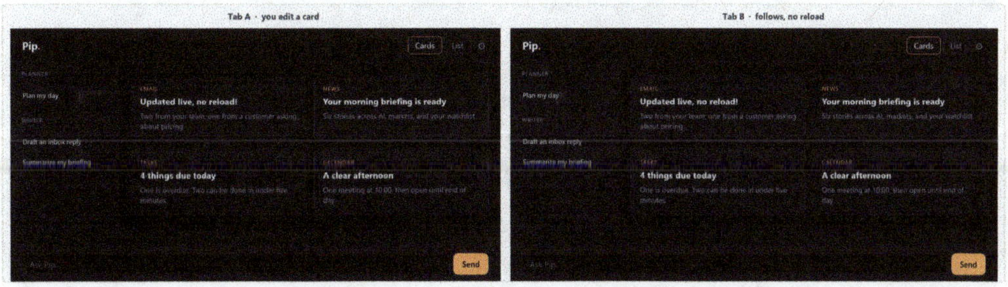

FIGURE 10.3 *One card edit, two tabs. Tab B updates itself the instant the change is announced, no reload.*

Surgical updates, not reloads

Notice we didn't call `location.reload()` when a card changed. We found the one card element by its slug and rewrote just its title and body. That precision matters more than it looks:

- A full reload would throw away everything the user has going: the open modal, scroll position, the chat conversation, any local state. A surgical update leaves all of that untouched and changes only what actually changed.
- It's faster and calmer: one element updates with a gentle flash, instead of the whole page blinking.

The pattern generalizes: *identify the smallest DOM element that reflects the changed data, and update only that.* Cards update by slug; badges toggle by agent name. Each event type knows exactly which element it owns. When you add a new live-updating thing, the question to answer is "what one element does this change, and how do I find it?" and the answer is almost always "by an id or a `data-` attribute," which is why we made everything addressable back in Chapter 3.

NORTH STAR · CORA

Cora's live pipeline is this same shape with an industrial front end. Instead of publishing from application code, its database does the announcing: a Postgres trigger fires `NOTIFY` on every insert, update, or delete to the watched tables, so *any* change, from the app, a background agent, or a hand-edited row, flows through automatically. A background thread holds a long-lived `LISTEN` connection, forwards each notification to the async loop, and a `ChangeBroadcaster` fans it out to one bounded queue per browser (dropping the oldest on overflow, exactly like Pip). The browser **debounces** a burst of events over 250ms into a single re-fetch, so ten rapid updates become one, then does surgical DOM updates: replacing a card's text in place, syncing the matching bubble, flashing the changed element with an orange halo. Cora's agent tabs light up through this pipeline, which is why they glow for background work no browser started. Pip's `broadcaster.publish` is the hand-rolled version of Postgres `NOTIFY`; everything downstream is the same idea.

The gotcha: a real-world wart, and the honest limits

Every production push system has warts. Here are the ones worth knowing, because they're the difference between "works on my machine" and "works."

The threading wart. Our skill code runs in a *worker thread* (FastAPI runs sync `def` endpoints in a threadpool so they don't block the async loop). But the SSE queues live on the *async event loop*. You cannot just call `queue.put_nowait` from the worker thread, because asyncio objects aren't thread-safe, and doing so leads to events that silently never arrive, or occasional crashes. That's exactly why `publish` uses `loop.call_soon_threadsafe`: it's the one correct bridge from a thread into the loop. This is Pip's version of a wart Cora hit hard: on Windows, the async database driver refuses uvicorn's default event loop entirely, so Cora runs its `LISTEN` connection in a **plain sync thread** and bridges into the loop the same way, side-stepping the whole problem. When your push channel "works sometimes," suspect a thread/loop boundary before anything else.

No replay: events during a gap are lost. SSE has no memory. If a browser is disconnected for two seconds (a network blip, a laptop sleeping) and three events fire in that window, those three events are simply gone. The browser reconnects and only hears what happens *next*. This is usually fine because each update is idempotent and the *next* event corrects the screen. But it means push is a fast path, not a guarantee. The safety net is a plain periodic refresh (or the user's own reload). Don't treat SSE as a durable log; treat it as a low-latency nudge, with a coarser mechanism underneath for correctness.

Bounded queues, dropped on overflow. We capped each queue at 100 and drop on full. That's deliberate: a browser that stops reading (a background tab throttled by the OS) must not accumulate an unbounded backlog on the server. Dropping is safe precisely *because* updates are idempotent. A missed event just means that browser is briefly stale until the next event or refresh sets it right. An unbounded queue is a memory leak waiting for one slow client.

The through-line:

> **Push is a fast, best-effort nudge, not a durable guarantee. Make every update idempotent, bound your queues, bridge threads correctly, and keep a coarse refresh as the safety net.**

What you built

- A `Broadcaster` that fans change events out to every connected browser, thread-safely.
- A `/api/events` SSE endpoint with reconnection and heartbeats.
- `publish` calls on agent runs and a new `PATCH /api/cards/{slug}` endpoint.
- A browser listener that applies **surgical** updates to badges and single cards, never a full reload.
- Agent badges that now light up in every tab for any run, and cards that update live with a flash.
- The judgment to treat push as best-effort, keep updates idempotent, and bound your queues.

Exercises

1. **Prove it's cross-client.** Open three tabs. Trigger a skill in one and a card edit (the card-edit command above) in another, and watch the third tab reflect both without you touching it. You're seeing a single source of truth broadcast to many screens, the payoff of every "state in one place" decision so far.
2. **Add debounce.** Fire the card-edit command above five times in a row, fast. Notice the browser re-fetches `/api/cards` five times. Add a 250ms debounce around the card handler (collect slugs, then apply once after a short pause), Cora's exact optimization, and watch five bursts become one re-fetch. You've just felt why the debounce exists.
3. **Make the AI edit a card live.** Add an `update_card` action to the vocabulary that `PATCH`es `/api/cards/{slug}`, wire it in `executeAction`, and tell Pip "rename my inbox card to 'Urgent'." The AI writes the row, the server broadcasts, and the card updates itself in front of you: the writable surface (Chapter 4), the action protocol (Chapter 7), and live updates (this chapter) all firing at once.

Checklist before Chapter 11

- ☐ A skill run in one tab lights the agent badge in another tab.
- ☐ The card-edit command above updates and flashes the card in all open tabs, no reload.
- ☐ You can explain why `publish` uses `call_soon_threadsafe`.
- ☐ You understand that a dropped event is benign because updates are idempotent.
- ☐ Commit: `git add . && git commit -m "Chapter 10: live updates via SSE, push and surgical DOM updates"`

Pip now moves on its own. The next chapter is the optional capstone that turns it from a website into a *presence*: Chapter 11 adds a wake word and a living orb, so Pip can listen for its name and breathe while it speaks.

CHAPTER ELEVEN

Voice & Presence

> *This chapter is the optional capstone. Everything before it made Pip capable; this makes it feel like someone in the room. We add a wake word, so you can say "Hey Pip" and it listens, and a living orb that breathes while idle and stirs while it speaks. None of it is load-bearing: skip the whole chapter and Pip still works. But it's the moment the demo makes people lean in.*

There's a difference between an app you operate and a presence you talk to, and it's mostly theatre: a voice, and something that looks alive. This chapter is that theatre, built honestly. It leans on browser features that vary by browser (voice recognition is best in Chrome and Edge), so treat it as a bonus layer that degrades gracefully: if the features aren't there, Pip quietly falls back to typing, and nothing breaks.

> **Voice is where an AIOS stops being a website and becomes a presence. It's optional, but it's the moment the demo makes people lean in.**

The seam

Two independent pieces:

- **Voice.** The browser can *hear* (speech-to-text) and *speak* (text-to-speech), both built in, both free. We'll wire them to the chat pipe you already have: heard speech becomes a message; Pip's reply gets spoken. And a **wake word** listens continuously for "Hey Pip" so you don't have to click anything.
- **A living orb.** A visual driven by a little animation loop: a gentle "heartbeat" at rest, a livelier envelope while it's listening or speaking. Crucially, we build in the exact spot where a *real* audio amplitude would plug in, so the orb could one day breathe with Pip's actual voice.

Neither touches the AI's logic. Voice is just another way to send a message and hear the reply; the orb is pure presentation. That separation is deliberate, because the brain doesn't care how words arrive.

Build the dumb version

Step 1: make chat callable from anywhere

Voice needs to send a message the same way the chat box does. So first, extract the send logic into one function both can call. In `shell.js`, replace your chat submit handler with this reusable `sendMessage(text)` that returns Pip's clean reply:

```javascript
// --- Chat (now callable from voice too) ----------------------------
----
async function sendMessage(text) {
  text = (text || '').trim();
  if (!text) return '';

  addBubble('user', text);
  history.push({ role: 'user', content: text });
  const pending = addBubble('pip', '…');

  try {
    const res = await fetch('/api/chat', {
      method: 'POST',
```

```
        headers: { 'Content-Type': 'application/json' },
        body: JSON.stringify({ messages: history, screen:
collectScreenContext() }),
      });
      const data = await res.json();
      const clean = runActionsAndClean(data.reply);
      pending.textContent = clean || '(done)';
      history.push({ role: 'assistant', content: clean });
      return clean;                          // ← voice will speak this
    } catch (err) {
      pending.textContent = '(Pip could not reach its brain.)';
      return '';
    }
  }

  chatForm.addEventListener('submit', (e) => {
    e.preventDefault();
    const text = chatText.value;
    chatText.value = '';
    sendMessage(text);
  });
```

Nothing about typing changed, but now `sendMessage` is a door that voice can walk through too.

Step 2: the orb

Give Pip a face. Add an orb element to `index.html`, just before the chat dock:

```html
<div class="orb" id="orb" title="Click to talk to Pip"></div>
```

Style it as a warm glowing sphere that sits above the chat:

```css
    .orb { position:fixed; right:24px; bottom:180px; width:64px;
height:64px;
           border-radius:50%; cursor:pointer; z-index:50; will-
change:transform;
           background:radial-gradient(circle at 35% 30%, #ffd9a8, var(--
accent) 60%, #7a3e08);
           box-shadow:0 0 30px rgba(251,146,60,.5); transition:box-
shadow .2s; }
    html[data-focus="voice"] .orb { box-shadow:0 0 40px
rgba(251,146,60,.9); }
```

Now drive it. This animation loop runs forever, scaling the orb from an amplitude value. At rest it's a soft two-beat heartbeat; when listening or speaking, it's a busier envelope. Add to `shell.js`:

```javascript
  // --- The living orb ---------------------------------------------
----
  const orb = document.getElementById('orb');
  let orbState = 'idle';    // 'idle' | 'listening' | 'speaking'

  function driveOrb(now) {
    const t = now / 1000;
    let amp;
    if (orbState === 'idle') {
      // Two Gaussian peaks per 1.6s: a slow "lub-dub" heartbeat.
      const p = t % 1.6;
      const beat = Math.exp(-((p - 0.2) ** 2) / 0.01)
                 + 0.7 * Math.exp(-((p - 0.45) ** 2) / 0.01);
      amp = 0.02 + 0.04 * beat;
    } else {
      // A simulated speech envelope. *** This is where a real audio
amplitude
      // source would plug in *** Replace this line with a live mic/TTS
level.
      amp = 0.06 + 0.06 * Math.abs(Math.sin(t * 7) + 0.5 * Math.sin(t *
11));
    }
    orb.style.transform = `scale(${1 + amp})`;
    requestAnimationFrame(driveOrb);
  }
```

```
requestAnimationFrame(driveOrb);
```

Refresh and you'll see the orb gently breathing in the corner. That heartbeat is Pip idling: alive, waiting. The comment marks the one line you'd change to make it pulse with a real voice; today it's a convincing simulation.

Step 3: hear and speak

Now the voice itself. The browser's `SpeechRecognition` turns speech into text; `speechSynthesis` speaks text aloud. Add this to `shell.js`:

```javascript
// --- Voice: listen and speak ---------------------------------------
---
const SR = window.SpeechRecognition || window.webkitSpeechRecognition;
let listening = false;

function speak(text) {
  if (!window.speechSynthesis || !text) return;
  orbState = 'speaking';
  const utter = new SpeechSynthesisUtterance(text);
  utter.onend = () => { orbState = 'idle'; };
  window.speechSynthesis.speak(utter);
}

// Start a single listening session: capture one spoken command, send
it,
// speak the reply.
function startListening() {
  if (!SR || listening) return;
  listening = true;
  document.documentElement.dataset.focus = 'voice';    // orb brightens
  orbState = 'listening';
  if (wake) wake.stop();                                // release the
mic (see gotcha)

  const rec = new SR();
  rec.lang = 'en-US';
  rec.interimResults = false;
```

```
    rec.onresult = async (e) => {
      const text = e.results[0][0].transcript;
      const reply = await sendMessage(text);
      speak(reply);
    };
    rec.onend = () => {
      listening = false;
      delete document.documentElement.dataset.focus;
      if (orbState === 'listening') orbState = 'idle';
      // Resume the wake word a beat later, once the mic is free.
      setTimeout(() => { if (wakeEnabled()) startWakeWord(); }, 400);
    };
    rec.start();
  }

  orb.addEventListener('click', startListening);    // click the orb to
talk
```

Click the orb, allow the microphone when asked, and say "what's on my screen?" Pip hears you, answers in the chat, *and speaks the answer aloud*, while the orb stirs. You're now talking to your AIOS.

Step 4: the wake word

Clicking is fine, but "Hey Pip" is the magic. A second, *continuous* recognizer runs in the background listening only for the wake phrase; on a match, it hands off to `startListening`. It's off by default (always-on listening is a real privacy choice) and toggled in settings. Add to `shell.js`:

JAVASCRIPT

```javascript
// --- Wake word -----------------------------------------------
----
let wake = null;

function wakeEnabled() { return localStorage.getItem('pip-wake') ===
'on'; }

function startWakeWord() {
  if (!SR || listening || wake) return;
  wake = new SR();
  wake.continuous = true;
  wake.interimResults = true;
  wake.lang = 'en-US';

  wake.onresult = (e) => {
    const said = Array.from(e.results)
      .map(r => r[0].transcript).join(' ').toLowerCase();
    if (/hey (pip|pop|peep|pipp)/.test(said)) {   // tolerate
mishearings
      wake.stop();
      startListening();
    }
  };
  wake.onend = () => {                             // keep it alive
while enabled
    wake = null;
    if (wakeEnabled() && !listening) startWakeWord();
  };
  try { wake.start(); } catch (e) { /* already running */ }
}

function stopWakeWord() {
  const w = wake; wake = null;
  if (w) w.stop();
}

if (wakeEnabled()) startWakeWord();
```

Finally, a toggle so the user can turn it on. Add a checkbox to the settings modal in `index.html`:

```html
<label class="wake-row">
  <input type="checkbox" id="wakeToggle"> Listen for "Hey Pip"
</label>
```

And wire it in `shell.js` (near the settings code):

```javascript
const wakeToggle = document.getElementById('wakeToggle');
if (!SR) { wakeToggle.disabled = true; }        // browser can't do
it, so disable
wakeToggle.checked = wakeEnabled();
wakeToggle.addEventListener('change', () => {
  localStorage.setItem('pip-wake', wakeToggle.checked ? 'on' : 'off');
  if (wakeToggle.checked) startWakeWord(); else stopWakeWord();
});
```

Open ⚙ settings, tick "Listen for Hey Pip," grant the mic, and, from across the room, say "**Hey Pip, switch to list view.**" It wakes, hears the command, and (thanks to Chapter 7) *does* it. That's the whole system responding to your voice, hands-free.

The plug-in seam for a real voice

Look again at the one commented line in `driveOrb`:

```javascript
// *** This is where a real audio amplitude source would plug in
***
amp = 0.06 + 0.06 * Math.abs(...);    // simulated
```

Today the orb *fakes* its speaking motion with sine waves. To make it breathe with Pip's *actual* voice, you'd feed it a live amplitude: connect the `speechSynthesis` output (or, better, a real TTS audio stream) into a Web Audio `AnalyserNode`, read its level each frame, and set `amp` from that. The point of building it this way is that the seam is *already there*. The orb reads an `amp` variable and doesn't care whether the number is simulated or real. Swapping one for the other is a local change, not a rewrite. That's a habit worth generalizing: when you build a placeholder, build it behind the same interface the real thing will use, so the upgrade is a substitution.

NORTH STAR · CORA

Cora's voice and presence are this same design, more elaborate. Its wake word (`setupWakeWord`) runs browser-native `SpeechRecognition` continuously, off by default, matching `/hey (cora|core|cory|coda|corra)/`, the same tolerate-the-mishearings trick you used. Its orb (`driveSpeechOrb`) runs the same kind of loop: an idle `heartbeat(t)` of two Gaussian peaks per 1.6 seconds (Pip's heartbeat is lifted straight from it), easing into a simulated speech envelope when the mic is live, and its code carries the very same comment marking where a real TTS amplitude source plugs in. Cora even has *two* orbs, a foreground speech orb and a background WebGL orb, driven by separate loops. And it names the privacy trade-off explicitly, which we should too.

The gotcha: two things want the microphone, and a privacy debt

The mic hand-off. You have two recognizers: the wake-word listener (always on) and the command session (fires after "Hey Pip"). They both want the microphone, and the microphone has one owner at a time. If you don't hand it off cleanly, they fight: the command session can't start because the wake listener is holding the mic, or the wake listener restarts mid-command and swallows it. That's why `startListening` calls `wake.stop()` first, and why `rec.onend` waits 400ms before resuming the wake word, giving the OS a beat to fully release the mic before the other recognizer grabs it. Any always-listening feature that also needs the mic on demand has this problem; the fix is always an explicit stop-then-start with a small gap, never two things assuming they can share.

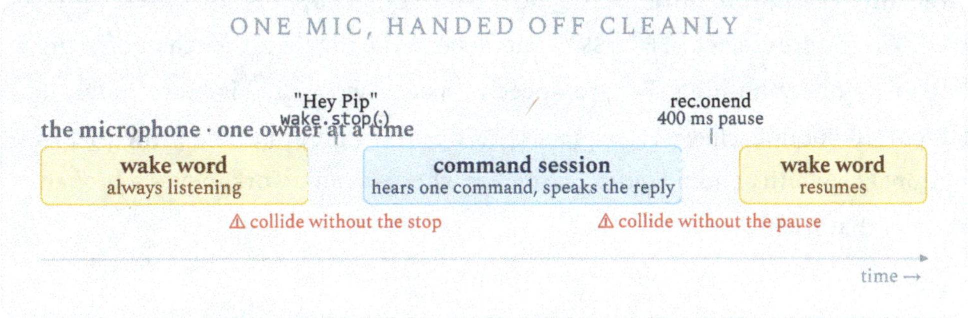

FIGURE 11.1 *Two recognizers, one mic. Each hand-off is an explicit stop, a gap, then a start, never a shared grab.*

> **When two features want the same microphone, hand it off explicitly: stop one, pause, start the other. They cannot share it, and assuming they can makes both flaky.**

The privacy debt, and saying it out loud to your users. In Chrome, the browser's `SpeechRecognition` doesn't transcribe on your device. It streams short audio buffers to Google's servers and gets text back. That means an always-on wake word is, technically, sending ambient audio off the machine whenever it's listening. For a personal demo that's often acceptable; as a default that's silently on, it is not. This is exactly why Pip's wake word ships **off by default** and behind an explicit toggle the user opts into, because the user should choose to be listened to, knowingly. If you want a genuinely local wake word, you'd swap `SpeechRecognition` for an on-device engine (a WASM wake-word detector like Picovoice Porcupine), which never sends audio anywhere. Cora notes the same trade-off and the same escape hatch. The lesson generalizes past voice: **any feature that sends user data somewhere should be opt-in and honest about it**, not a convenience quietly enabled.

Graceful degradation. Firefox and older Safari don't support `SpeechRecognition` well. We feature-detect (`if (!SR)`) and disable the toggle rather than offering a button that does nothing. Text-to-speech (`speechSynthesis`) is more widely supported but not universal, so `speak` no-ops if it's missing. The principle: an optional capability should *quietly be absent* where it can't work, never a broken control that frustrates.

What you built

- A reusable `sendMessage(text)` that both typing and voice flow through.
- A living orb driven by an animation loop, with an idle heartbeat and an active envelope, and the real-audio plug-in seam built in.
- Click-to-talk and a "Hey Pip" wake word (off by default, opt-in), wired to the existing chat and action pipeline.
- Spoken replies via the browser's text-to-speech.
- A clean microphone hand-off, honest privacy defaults, and graceful degradation where the browser can't help.

Exercises

1. **Change the wake phrase.** Make Pip answer to a name you like by editing the regex in `startWakeWord`. Say it a few times and tune the alternatives to match how your browser mishears it. You'll feel *why* the pattern is deliberately loose.
2. **Drive the orb from a real level.** As a stretch: create a Web Audio `AudioContext`, an `AnalyserNode`, and (using `getUserMedia`) feed the microphone into it; read `analyser.getByteFrequencyData` each frame and set the orb's `amp` from the average while listening. Now the orb reacts to *your actual voice level*, and you've replaced the simulated line with a real one, exactly at the seam the code was built around.
3. **Make degradation visible.** Open Pip in Firefox (or disable `SpeechRecognition`). Confirm the wake toggle is disabled and typing still works perfectly. Then decide: should the orb still be clickable there, or should it hint that voice is unavailable? Implement your choice. You're practicing the judgment of how an optional feature should *absent itself*.

Checklist before Chapter 12

- [ ] The orb breathes at rest and stirs while listening or speaking.
- [ ] Clicking the orb lets you speak a command; Pip answers aloud.
- [ ] "Hey Pip" (when enabled) wakes it and runs the spoken command.
- [ ] The wake word is off by default and behind an explicit toggle.
- [ ] In a browser without speech support, nothing is broken and the toggle is just disabled.
- [ ] Commit: `git add . && git commit -m "Chapter 11: voice + a living orb (optional capstone)"`

Pip can now be seen, spoken to, and heard from. One chapter remains, and it isn't a feature but the discipline that ties everything together. Chapter 12 is about *running, debugging, and iterating,* including the honest truth that your test suite is a pair of eyes, and the habits that keep a fast-changing AIOS from becoming a mess.

CHAPTER TWELVE

Run, Debug, Iterate

> *The last build chapter isn't a feature. It's the discipline that keeps everything you've built from rotting. An AIOS has no unit test that says "it feels alive," so your verification loop is a pair of eyes. This chapter makes that loop fast, hands you a debugging playbook for the silent failures the whole book warned about, and teaches the habit of reversible changes that lets you move fast without making a mess.*

You've built a lot. Pip sees, acts, remembers, runs skills, updates itself, and talks. From here on you'll mostly be *changing* it: adding a skill, tweaking a prompt, trying an idea. And changing an AIOS is different from changing a normal app, because the thing you're really tuning is *behavior*, meaning how the model reasons over your context, whether an action fires, whether a reply lands right. There's no assertion for "that felt off." So the two skills that matter most now are: seeing what's wrong quickly, and being able to undo any experiment cleanly.

> **An AIOS has no unit test that says "it feels alive." Your verification loop is looking at it. Make that loop fast.**

The seam

The whole loop is: change something → look at it → decide. Everything in this chapter shortens the middle step. A fast look means the server is already running, the page is one refresh away, and when something's wrong you can *poke the system by hand* to find out why instead of guessing. Build that and iteration becomes pleasant; skip it and every change is a slow, uncertain slog.

Build the dumb version

One command to run

You already have it, and it's worth appreciating how little it is:

```bash
python -m uvicorn server:app --reload --port 8000
```

One process serves the page *and* the API. `--reload` restarts the server automatically whenever you save a Python file. So your loop is: edit, save, refresh the browser. Keep this running in its own terminal all day; you rarely stop it. (Frontend files like `index.html`, `shell.js`, and `bootstrap.js` don't even need the restart; a browser refresh picks them up.)

Devtools hooks: poke the system by hand

Here's the highest-leverage debugging tool you have, and it costs a few lines: **expose functions on `window` so you can drive Pip manually from the browser console.** When something the AI is *supposed* to trigger doesn't work, you find out fast by triggering it yourself.

Expand the `window.pip` object at the bottom of the `shell.js` IIFE into a proper debug kit:

```javascript
  // --- Debug hooks: drive Pip by hand from the DevTools console ----------
  window.pip = {
    openCard,                                  // pip.openCard('inbox')
    runSkill,                                  // pip.runSkill('plan-day')
    sendMessage,                               // pip.sendMessage('switch to list view')
    setView,                                   // pip.setView('list')
    refresh: () => location.reload(),          // pip.refresh()
    forceOrb: (state) => { orbState = state; }, // pip.forceOrb('speaking')
```

```
    notes: () => fetch('/api/persona').then(r =>
r.json()).then(console.table),
  };
```

Now open DevTools (F12) and try it. `pip.setView('list')` changes the layout. `pip.openCard('tasks')` opens the modal. `pip.sendMessage('open my briefing')` runs a full chat turn, action and all, as if you'd typed it. `pip.notes()` prints a table of what Pip remembers. You've turned every internal capability into something you can fire on demand, which is exactly how you isolate a bug: *does the function work when I call it directly?* If yes, the problem is upstream (the AI didn't emit the action, or the parser missed it). If no, the problem is the function itself. That one question, answerable in a second, cuts most debugging in half.

And you already have a passive hook: the `[pip-action]` console logs from Chapter 7. Keep DevTools open while you work, because those lines tell you exactly what the AI's hands did, every turn.

The honest truth about testing

Pip has no test suite. No linter. No build step. The verification loop is you, looking at a browser. That isn't a gap this book forgot to fill. It's the honest state of a project this size, and pretending otherwise would set you up to fail.

Say the same thing in your own project's README. It's tempting to imply more rigor than exists, but a new contributor who's told "run the tests" and finds none wastes an hour looking. "There is no automated test suite; verification is visual inspection in a browser, and here's the checklist" is more useful and more honest. As Pip grows, you'll *want* tests around the pure logic (the action parser is a great candidate, since it's a pure function from text to a list of actions), and adding them is worthwhile. But the *behavior*, whether Pip feels right, will always come down to using it. Make peace with that, and make the looking fast.

The debugging playbook

Almost every bug in this book was from the same family: **it does nothing, and throws no error.** That silence is the signature of an AIOS bug, and it's why guessing is so tempting and so wasteful. Here's a playbook that maps the symptom straight to the cause. Keep it near.

<table>
<thead>
<tr><th>SYMPTOM</th><th>MOST LIKELY CAUSE</th><th>HOW TO CHECK</th></tr>
</thead>
<tbody>
<tr><td>A feature does nothing, no console error</td><td>Load order: behavior ran before the element existed (Ch 3)</td><td>Confirm <code>bootstrap.js</code> loads <code>shell.js</code> last, after rendering. In the console, does <code>document.querySelector('.card')</code> find anything?</td></tr>
<tr><td>Cards render empty</td><td>The database is empty, because you didn't run <code>init_db.py</code> (Ch 4)</td><td>Does <code>pip.db</code> exist? Run <code>python init_db.py</code>. Hit <code>/api/cards</code> directly. Is the array empty?</td></tr>
<tr><td>Pip ignores your instructions, acts generic</td><td>The system prompt isn't reaching the model (Ch 5)</td><td>Run the "~ Pip 🐦" sign-off probe. If it doesn't appear, the prompt is being dropped, so fix delivery rather than wording.</td></tr>
<tr><td>Pip says it did something, but nothing happened</td><td>The action block didn't parse, from a sloppy fence (Ch 7)</td><td>Watch the <code>[pip-action]</code> logs. A parse warning means malformed output; no log means the model emitted nothing. Different fixes.</td></tr>
<tr><td>A live update never arrives</td><td>SSE gap, no subscriber, or a thread/loop slip (Ch 10)</td><td>Is the <code>EventSource</code> connection open in the Network tab? Does <code>publish</code> use <code>call_soon_threadsafe</code>? A refresh should fix a one-off (updates are idempotent).</td></tr>
<tr><td>The server hangs at "Reloading..."</td><td>Background tasks and <code>--reload</code> don't mix (below)</td><td>Kill every server process and restart fresh; see the gotcha.</td></tr>
<tr><td>Voice does nothing</td><td>Browser doesn't support it, or mic not granted (Ch 11)</td><td>Is <code>window.SpeechRecognition || window.webkitSpeechRecognition</code> defined? Check the site's mic permission.</td></tr>
</tbody>
</table>

The meta-lesson: when a feature fails *silently*, don't stare at the code. Ask "which layer went quiet?" and use the hooks and logs to find out. Almost every entry above is answered by a single console command or a glance at DevTools.

The habit that keeps fast from becoming reckless

An AIOS changes fast, which is its charm and its danger. The thing that lets you experiment freely without fear is a clean **undo**: the confidence that any change can be reversed cleanly, in one command. That confidence comes from a small, boring workflow you should adopt now.

One change, one branch, one changelog entry, one merge. Concretely, for any real change to Pip:

```bash
git checkout -b pip_theme_toggle        # 1. branch off main, named for the change
# ... make the change ...
# 2. add an entry to the TOP of CHANGELOG.md (see below)
git add . && git commit -m "Add a light/dark theme toggle"   # 3. commit on the branch
git checkout main
git merge --no-ff pip_theme_toggle      # 4. merge, keeping the branch boundary visible
```

Create a `CHANGELOG.md` and put a dated entry at the top of it for each change, including the load-bearing part, a concrete "**To revert**" line:

```markdown
## Light/dark theme toggle

Date: 2026-08-14
What changed: added a theme toggle to the header; theme stored in localStorage and applied in the pre-paint script. Files: index.html, shell.js.
To revert: git revert -m 1 <the merge commit hash>
```

Two things make this pay off. The `--no-ff` merge keeps each change as a visible unit in `git log --graph`, instead of a flat smear of commits, so you can *see* where one experiment starts and ends. And because each change is one merge, reverting it is exactly one command: `git revert -m 1 <merge-hash>`, straight from the changelog entry. When an experiment turns out badly, whether a prompt tweak that made Pip worse or a feature that didn't land, you undo it cleanly and move on, instead of picking apart tangled commits. Fast iteration is only safe when reversal is cheap. This is how you make it cheap.

(Trivial one-line fixes can land on `main` directly; it's a judgment call. The workflow is for anything you might want to cleanly undo.)

NORTH STAR · CORA

Cora runs and iterates exactly this way. One command (`python -m uvicorn server:app --reload --port 8000`) serves everything. It exposes the same kind of devtools hooks: `window.coraRefresh()` to force a data refresh, a dispatchable `cora:mic-toggle` event to simulate the mic, `document.documentElement.dataset.orbCenter = 'plexus'` to force a visual mode by hand, so any state is reachable from the console. It states plainly that there is **no test suite, no linter, and no build step, and that visual inspection in a browser is the verification loop.** And its whole workflow is the branch-and-changelog discipline you just learned: branch from `main` per change, a `CHANGELOG.md` entry with a `To revert:` line, one commit, merge with `--no-ff`, so reverting one task is a single `git revert` of the merge. Pip's process *is* Cora's process, and the discipline scales unchanged from a weekend project to a real system.

The gotcha: background tasks make hot-reload flaky

Here's the one that'll bite you now that Pip has a live-update channel: **`--reload` and background/streaming work don't always coexist peacefully.**

Cora hits this hard, and Pip can too. When there are long-lived connections (your SSE streams) and asynchronous work in flight, uvicorn's `--reload` sometimes hangs partway through a restart. You see "Reloading..." and then *nothing*, no "Started server process" line, and the page stops responding. The reloader is trying to tear down a process that's holding open streams, and it wedges.

There's no clever fix; there's a **ritual**, and knowing it saves you from thinking Pip is broken:

1. Stop the server: `Ctrl+C` in its terminal. If it doesn't die cleanly, force it:
   - **macOS / Linux:** `pkill -f "uvicorn server:app"`
   - **Windows (PowerShell):** find and stop the Python processes running uvicorn, or as a blunt instrument `taskkill /F /IM python.exe` (this kills *all* Python, which is fine on a dev machine where Pip is the only one running).
2. Start it again fresh: `python -m uvicorn server:app --reload --port 8000`.

A clean restart clears the wedged state every time. The lesson generalizes: once your app holds long-lived connections or runs background tasks, hot-reload becomes unreliable, and a **clean-restart ritual** is part of your normal toolkit rather than a sign something is wrong. Cora's team ran exactly this recipe "a half-dozen times this session"; it's routine, not alarming.

> **Once an AIOS holds live connections or background work, `--reload` gets flaky. Keep a clean-restart ritual, killing every server process and starting fresh, and don't mistake a wedged reload for a broken app.**

What you built (and learned)

- A one-command run loop, and the habit of leaving it running.
- A `window.pip` debug kit that lets you fire any internal capability from the console.
- The honest framing of testing an AIOS: the loop is your eyes, made fast, and where automated tests *do* belong (pure logic like the parser).
- A debugging playbook that maps silent failures straight to their cause.
- The branch → changelog → `--no-ff` workflow that makes any change reversible in one command.
- The clean-restart ritual for when live connections wedge hot-reload.

Exercises

1. **Debug with the hooks.** Deliberately break something small, such as misspelling a slug in a skill's seed or a card's `data-slug` in `bootstrap.js`, then use the playbook and `window.pip` to find it *without reading the diff.* Practice the reflex: "does the function work when I call it by hand?" You're training the loop this whole chapter is about.
2. **Do one change the disciplined way.** Add any small feature (a theme toggle, a new skill, a fourth view mode) using the full workflow: branch, change, CHANGELOG entry with a `To revert` line, commit, `--no-ff` merge. Then actually run `git revert -m 1 <hash>` and watch the change cleanly disappear. Now you *trust* the undo, which is the whole point.
3. **Write your project's honest README.** Draft the "how to run and verify" section for Pip the way this chapter argues for: one run command, the devtools hooks, and the plain statement that verification is visual with a checklist. Notice how much more useful honesty is than implied rigor.

Checklist

- ☐ `window.pip` in the console can drive cards, skills, view, and a full chat turn by hand.
- ☐ You can name, from the playbook, the cause of at least three "silent failure" symptoms.
- ☐ You made one change through the branch → CHANGELOG → `--no-ff` workflow and reverted it cleanly.
- ☐ You know the clean-restart ritual for a wedged `--reload`.
- ☐ Commit: `git add . && git commit -m "Chapter 12: debug hooks, playbook, and the reversible-change workflow"`

That's the last build chapter. Pip is complete: it sees, acts, knows itself, runs skills through agents, updates live, and speaks, and you can change it fearlessly. The **Appendix** ties a bow on it with a glossary of the terms we've used, the full file listing of the finished `pip/` project in one place, and honest directions for where to take it next.

Final Thoughts

You did it. Whatever you were when you opened this book, tech head or not, you are now someone who has built an AI Operating System with your own hands. Pip sees, acts, and knows itself, and there is not a single line of it you cannot name. That was the whole point. Not to hand you a framework that does the magic behind a curtain, but to prove something quieter and more durable: that these systems are not beyond you. They never were. They were just never explained at the right altitude, by someone willing to show you the gotchas along with the happy path.

That is the thing I most want you to carry out of here. Not the code, which you can always look up, but the feeling that you can build this. Because you just did.

Here is a small confession. When I built Pip a second time, following the very chapters you just read, I was stunned at how fast it went. The original version, the one that became Cora, took me a long and winding road of dead ends, notes, and afternoons lost to silent failures. Those 154 sessions are where this book came from. What took me months of fumbling, you did in a weekend of typing, because the book is the map I did not have. That compression, from a hard journey into a clear path, is the best gift I know how to give.

So do not stop here.

Local models are getting smaller, faster, and cheaper to run on your own machine almost by the month. The Pip you built is not a finished product; it is a foundation with your name on it. Bend it toward your actual life. Point it at your inbox, your notes, your quirks, the small daily frictions only you have. Take it to the next level and turn it into something that genuinely augments your day and gives you back a little time. That was always the real goal, more time for the things and the people that matter most.

And thank you. Truly.

Thank you for reading this book, for typing along, for trusting a stranger's mental model long enough to build your own. And thank you to everyone who, over the years, kept pushing me to realize that the knowledge rattling around in my head was worth something to other people, worth draining out of my notes and into these pages where it could finally be of use to someone other than me. You are the someone. This was always for you.

One last thing, and hold onto it as you build whatever comes next:

> **AI is here to augment humans, not to replace them. Keep that in mind, always.**

Now go build.

~ freddie

Appendix

> *The reference shelf: a glossary of the terms this book uses, the complete final source of the `pip/` project gathered in one place, and honest directions for where to take it next.*

A. Glossary

Action. A single, named thing the AI is allowed to do to the UI (`open_card` , `set_view_mode`). Actions form a *closed vocabulary*; the AI can compose them but not invent new ones. (Chapter 7.)

Action protocol. The agreed format the AI uses to request an action: a fenced `pip-action` block of JSON, parsed and executed by the frontend.

Agent. A named, specialized worker (Pip has Writer and Planner) that owns skills and leaves an audit trail in `agent_runs` . Its badge lights up while it works. (Chapter 8.)

AIOS (AI Operating System). Not a chatbot with a UI attached, but a UI an AI can *perceive and drive*: a shared workspace the human and the AI both read from and write to. (Chapter 1.)

Bootstrap. The frontend step (`bootstrap.js`) that fetches data, renders the DOM, then loads the behavior script. Runs *before* `shell.js` . (Chapter 3.)

Closed vocabulary. The safety property of control: the AI's actions are a fixed list, each mapping to something a user could already do, so the AI has no power the user doesn't. (Chapter 7.)

`common_data` . One generic lookup table holding every "list of allowed values" (card categories, view modes, agents), so new vocabulary is one `INSERT` , never a migration. (Chapter 4.)

Context injection. Assembling text (persona, screen state, memory, skills) and prepending it to every chat turn. The chat endpoint is a *context-injecting proxy*, not a passthrough. (Chapters 5 to 9.)

Idempotent. An operation safe to repeat (running `init_db.py` twice; applying a live update twice). Idempotency is why dropping an SSE event is benign. (Chapters 4, 10.)

Perception. Giving the AI a fresh, accurate description of the screen each turn so it *knows* rather than guesses. The first OS moment. (Chapter 6.)

Control. Giving the AI a safe way to *act* on the screen. The second OS moment. (Chapter 7.)

Self-awareness. The AI knowing its own remembered preferences and its own architecture. The third OS moment. (Chapter 9.)

Persona note. An append-only, capped preference the user tells Pip to remember; folded into the prompt every turn so Pip applies it. (Chapter 9.)

Screen context. The assembled block describing the current screen (view mode, visible cards, open card, available actions, skills), rebuilt every turn. (Chapter 6.)

Slug. A short, unique, URL-safe id (`inbox` , `plan-day`) that names a thing across the DOM, the database, and the AI. The shared address that makes everything drivable. (Chapters 3 and 4.)

SSE (Server-Sent Events). A one-way server-to-browser stream used to *push* changes so the screen updates without a reload. (Chapter 10.)

Surgical update. Updating only the one DOM element a change affects, instead of reloading the page, preserving modal, scroll, and chat state. (Chapter 10.)

Wake word. A continuously-listening recognizer that opens a voice session on a phrase ("Hey Pip"); off by default for privacy. (Chapter 11.)

B. The complete `pip/` project

Full source on GitHub — clone it, diff against it, or start from it:
github.com/fpokrzywa/aios_beginner_pip.git

Every file in its final form. This is the reference build; if your Pip misbehaves, diff against these.

```
pip/
├── .env.example        # template for .env (which is git-ignored)
├── .gitignore
├── requirements.txt
├── schema.sql          # database shape
├── db.py               # connection helper
├── init_db.py          # create + seed the database
├── llm.py              # swappable brain (Ollama / Anthropic)
├── server.py           # FastAPI: API + static frontend
├── index.html          # the whole page (structure + styles)
├── bootstrap.js        # fetch, render, SSE, then load shell.js
└── shell.js            # all behavior
```

`requirements.txt`

```
fastapi[standard]
uvicorn
httpx
python-dotenv
anthropic
```

C. Where to go next

You have a complete AIOS and, more importantly, the mental model behind it. Here's where the road continues, roughly easiest to hardest. Each one is a place the book deliberately stopped.

- **Native tool-calling.** Replace the text-fence action protocol with the Anthropic SDK's structured `tools` for the cloud brain, which gives more reliable parsing and no fence to mislabel. Keep the fence as the fallback for local models. (Chapter 7's "Layer 1.")
- **Test the pure logic.** `runActionsAndClean` is a pure function from text to actions, which makes it the perfect place for real unit tests. Feed it the malformed outputs from Chapter 7's gotcha and assert it recovers.
- **More agents, more skills.** You built the machinery once; adding an agent is now a prompt and a badge. Build a real one end-to-end (the discipline from Chapter 8): a "briefing" agent that fetches actual RSS, or a "scout" that watches a page.
- **Real data.** Pip's cards are seeded. Wire one card to something live, such as your actual calendar, a weather API, or an inbox, and watch perception and control suddenly matter more.
- **Debounced, trigger-driven live updates.** Add Cora's 250ms debounce (Chapter 10 exercise), and then, for the bigger leap, move from publishing in application code to database triggers, so *any* change to the data pushes automatically. On Postgres that's `NOTIFY`; the pattern is Cora's whole live pipeline.
- **Retrieval (RAG) for deep self-knowledge.** `BUILD_SUMMARY` is a fixed paragraph. For questions deeper than a summary can hold, such as "show me the code that parses actions," index your own source files and retrieve the relevant piece into the prompt. This is Cora's noted next step, and the natural ceiling of Chapter 9.
- **Auth and multi-user.** Pip is single-user. Adding accounts (sessions, hashed passwords) and scoping every table by user is the step from "my demo" to "a product." It also forces the security questions the book only flagged: sanitizing card HTML, guarding against prompt injection in anything user- or

- web-sourced, and making sure only the user can write persona notes.
- **A better brain seam.** Give each agent its own model config (Cora's `agent_configs`), so Writer can run on a cheap local model while a heavy planning agent uses a cloud one, all behind the `call_llm` boundary you already built.

The through-line for all of it is the rule from Chapter 1: **keep state in one place both the UI and the AI can read.** Every feature above is another thing to perceive, another thing to drive, another section of context, the same three moves you've now made a dozen times. That's the quiet promise of an AIOS built this way: it grows by addition, not by rewrite.

You started with an empty folder. You built a system that can see its own screen, act on it, remember you, and explain itself, and you understand every line. Now go make it yours.